Survival Guide

General Chemistry with Math Review

THIRD EDITION

Charles H. Atwood
University of Georgia

Prepared by

Charles H. Atwood
University of Georgia

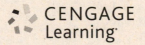

CENGAGE
Learning

Australia • Brazil • Mexico • Singapore • United Kingdom • United States

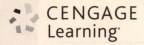

For product information and technology assistance, contact us at **Cengage Learning Customer & Sales Support, 1-800-354-9706**.

For permission to use material from this text or product, submit all requests online at **www.cengage.com/permissions** Further permissions questions can be emailed to **permissionrequest@cengage.com**.

Cover Image: ©Lightspring/Shutterstock

ISBN: 978-1-305-62956-1

Cengage Learning
20 Channel Center Street
Boston, MA 02210
USA

Cengage Learning is a leading provider of customized learning solutions with office locations around the globe, including Singapore, the United Kingdom, Australia, Mexico, Brazil, and Japan. Locate your local office at: **www.cengage.com/global**.

Cengage Learning products are represented in Canada by Nelson Education, Ltd.

To learn more about Cengage Learning Solutions, visit **www.cengage.com**.

Purchase any of our products at your local college store or at our preferred online store **www.cengagebrain.com**.

Printed in the United States of America
Print Number: 01   Print Year: 2016

Table of Contents

Module 4
The Mole Concept

Module 5
Chemical Reaction Stoichiometry

Module 6
Chemical Reaction Types

Module 7
Electronic Structure of Atoms

Module 13
States of Matter

Module 14
Solutions

Module 15
Heat Transfer, Calorimetry, and Thermodynamics

Module 16
Chemical Kinetics

Module 17
Gas Phase Equilibria

Module 18
Aqueous Equilibria

Module 19
Electrochemistry

Module 20
Nuclear Chemistry

Practice Test Six

Math Review

Practice Test Solutions

Preface

To The Student

In the third edition of the **Survival Guide for General Chemistry with Math Review** I have tried to improve upon the first and second editions by adding more help for you the student. In the first edition, I tried to write it as if you were sitting beside me at my desk and I was helping you solve problems. Consequently, in this edition solved problems with numerous arrows and boxes pointing where numbers, equations, and other pertinent information come from and are introduced into the problems still remain. Many general chemistry course commonly asked questions have been distilled into a small set of topics in this edition. Also left from the first edition are the **INSIGHT** and **CAUTION** boxes drawing your attention to important details.

In the second edition, I included **TIPS** boxes where a collection of several ideas pertaining to that topic help you distinguish a problem from other similar problems allowing you to categorize each problem. I also included **Module Predictor Questions**, with worked out solutions, and **Practice Tests** designed to test your knowledge over several Modules. Each question in the **Module Predictor Questions** has an indicated proficiency level (1 = basic, 2 = mid-level, and 3 = high) next to it. I suggest you attempt the **Module Predictor Questions** before tackling the Module. If you cannot work the level 2 and 3 questions, it is best that you review that Module improving your understanding of those topics. Approximately every five modules, a **Practice Test** with predictor questions is included to give you a measure of how well you comprehend the previous modules. Hopefully, this will assist in studying for tests at your institution.

In the third edition I did an extensive rewrite of many problems. Every effort was made to fix mistakes, remove errors, and clean up the problems. At the end of the initial module and up to module 6 I have placed a series of study tips designed to help students improve their studying skills. If you heed these tips, they are known to improve student success. Also at the end of every module I have added some graphics designed to show you how each module connects to the following modules. This is an effort to help you integrate your knowledge from module to module. I hope that these new additions will aid your comprehension and success in general chemistry.

Acknowledgements

Once again, my Cengage Learning colleagues, Lisa Lockwood and Peter McGahey, have been instrumental in this **Survival Guide** revision. Lisa's advice and guidance insured that the improvements are even more usable for the students. Peter handled the details of getting the proofs into production. I am fortunate to have such friends and colleagues.

My two children, Louis and Lesley, are always in my thoughts. Both are tackling the world on their own terms. I hope that you continue to grow and experience the wonderful lives you have set out upon. I love you both.

My wife, Judy, watches me disappear to my home office nearly every evening to work on this and other projects. She never ceases to amaze me with her love for me and her willingness to let me work on various projects at the expense of her time with me. I cannot adequately express in words my love and appreciation to her. I dedicate this edition of the **Survival Guide** to her.

Module 1 Predictor Questions

The following questions may help you determine the extent you need to study this module. Questions are ranked according to ability.

        Level 1 = basic proficiency
        Level 2 = mid level proficiency
        Level 3 = high proficiency

If you can correctly answer Level 3 questions you probably do not need to spend much time on this module. If you can only answer Level 1 problems, you should review this module.

Level 2   1. How many Mm are there in 427 miles?

Level 2   2. Convert 1.52×10^4 cm^3 to ft^3.

Level 2   3. Determine the number of significant digits in each of the following numbers.
            a) 3700
            b) 770.
            c) 770.0
            d) 0.00420
            e) 8.12×10^4

Level 1   4. Answer this addition problem using the correct number of significant figures: $101.22 + 222.3 =$

Level 1   5. Answer this multiplication problem using the correct number of significant figures: $8.4 \times 8.22 =$

Level 1   6. What is the answer to this numerical calculation, using the correct number of significant digits? $(67.888 - 7.64 + (1.2 \times 10^2)) / 3.27 =$

Level 1   7. A cubic sample of iron has an edge length of 6.32 in. The density of iron is 7.86 g/cm^3. What is the mass of this iron sample?

Module 1 Predictor Question Solutions

1. How many Mm are there in 427 miles?

$$427 \; \text{miles} \left(\frac{5280 \; \text{ft}}{1 \; \text{mile}} \right) \left(\frac{12 \; \text{in}}{1 \; \text{ft}} \right) \left(\frac{2.54 \; \text{cm}}{1 \; \text{in}} \right) \left(\frac{1 \; \text{m}}{100 \; \text{cm}} \right) \left(\frac{1 \; \text{Mm}}{1 \times 10^6 \; \text{m}} \right) = 0.687 \; \text{Mm}$$

2. Convert $1.52 \; x \; 10^4$ cm^3 to ft^3.

$$1.52 \times 10^4 \; \text{cm}^3 \left(\frac{1 \; \text{in}}{2.54 \; \text{cm}} \right)^3 \left(\frac{1 \; \text{ft}}{12 \; \text{in}} \right)^3 =$$

$$1.52 \times 10^4 \; \text{cm}^3 \left(\frac{1 \; \text{in}^3}{16.39 \; \text{cm}^3} \right) \left(\frac{1 \; \text{ft}^3}{1728 \; \text{in}^3} \right) = 0.537 \; \text{ft}^3$$

3. Determine the number of significant digits in each of the following numbers.
   a) 3700

   2 significant figures the trailing zeros are not significant

   b) 770.

   3 significant figures significant figures; the trailing zero is significant due to the decimal

   c)   770.0   **4 significant figures; the trailing zeros are significant due to the decimal**

   d) 0.00420 **3 significant figures; only the last zero is significant**

   e) $8.12 \; x \; 10^4$ **3 significant figures; the digits in 10^4 are not significant**

4. Answer this addition problem using the correct number of significant figures:
   $101.22 + 222.3 =$ **323.5 The answer is limited to one decimal place.**

$$101.22$$
$$\underline{+ \; 222.3}$$
$$323.5$$

5. Answer this multiplication problem using the correct number of significant figures: $8.4 \; x \; 8.22 =$ **69 The answer is limited to two significant figures.**

6. What is the answer to this numerical calculation, using the correct number of significant digits? $(67.888 - 7.64 + (1.2 \; x \; 10^2)) / 3.27 =$ **55**
The subtraction step is limited to two decimal places (four significant figures). The addition step is limited to one decimal place (two significant figures) because the first place that two numbers have a significant figure in

2

common is in the tens place. The division step is then limited to two significant figures. The answer is 55.

7. A cubic sample of iron has an edge length of 6.32 in. The density of iron is 7.86 g/cm^3. What is the mass of this iron sample?

$$6.32 \text{ in} \left(\frac{2.54 \text{ cm}}{1 \text{ in}} \right) = 16.1 \text{ cm}$$

$$\text{Volume} = \left(16.1 \text{ cm} \right)^3 = 4.17 \times 10^3 \text{ cm}^3$$

$$\text{Mass} = \left(4.17 \times 10^3 \text{ cm}^3 \right) \left(\frac{7.86 \text{ g}}{\text{cm}^3} \right) = 3.28 \times 10^4 \text{ g}$$

Study Tip #1

Experts in every field have a common problem solving approach that is outlined below. If you use these steps when tackling chemistry problems it will help your problem solving ability.

1. Recognize the Problem

    After looking at the problem ask yourself, "What's this problem asking and what do I need for an answer?"

2. Describe the problem in terms of the field

    What does this problem have to do with what I have seen in class or in the textbook?

3. Plan a solution

    How do I start with what I have been given to get a result that addresses the problem?

4. Execute the plan

    Go through a systematic process to get an answer.

5. Evaluate the solution

    Does my answer make sense? Can this answer be true?

This process is not a linear sequence. It requires continuous reflection and iteration. If your answer does not make sense, repeat the process to understand where you have gone astray.

Personal communication: K. Heller

Module 1
Metric System, Significant Figures,
Dimensional Analysis, and Density

Introduction
This module addresses several topics typically introduced in the first chapter of general chemistry textbooks. This module describes:
1. the basic rules of the metric system and significant figures
2. how to use dimensional analysis to solve problems
3. the relationship between density, mass, and volume and how to apply dimensional analysis to density problems.

Module 1 Key Equations & Concepts

1. $\text{density} = \dfrac{\text{mass}}{\text{volume}} = \dfrac{m}{v}$

   If any two of the variables in the equation are known, you can solve for the third using basic algebra.

   The metric system uses a series of multipliers to convert from one size unit to another. You must be very familiar with these prefixes and how to convert between size units. A common set of multiplier prefixes is given in the table below.

These prefixes may be used with any unit of measurement and the relationship between the *base unit* and the unit with the prefix is the same regardless of the base unit. The base unit is represented by *x* in the table.

Pay special attention to the unit factors provided as they are used to convert one unit to another. Note that each unit factor may be written in two equivalent ways. You choose the one to use depending on the units you are trying to cancel in the dimensional analysis problem (see examples below).

One way to help ensure you work conversion problems correctly is to remember which unit is the largest. For example, if you are converting from pg to Mg, then keep in mind that a Mg is much, much larger than a pg. So, the numerical value should get much smaller as you convert from pg to Mg.

Prefix Name	Prefix Symbol	Multiplication Factor	Unit Factors
mega-	M	1000000 or 10^6	$\dfrac{1\,Mx}{10^6\,x} = \dfrac{10^6\,x}{1\,Mx}$
kilo-	k	1000 or 10^3	$\dfrac{1\,kx}{10^3\,x} = \dfrac{10^3\,x}{1\,kx}$
deci-	d	0.1 or 10^{-1}	$\dfrac{1\,dx}{10^{-1}\,x} = \dfrac{10^{-1}\,x}{1\,dx}$
centi-	c	0.01 or 10^{-2}	$\dfrac{1\,cx}{10^{-2}\,x} = \dfrac{10^{-2}\,x}{1\,cx}$
milli-	m	0.001 or 10^{-3}	$\dfrac{1\,mx}{10^{-3}\,x} = \dfrac{10^{-3}\,x}{1\,mx}$
micro-	μ	0.000001 or 10^{-6}	$\dfrac{1\,\mu x}{10^{-6}\,x} = \dfrac{10^{-6}\,x}{1\,\mu x}$
nano-	n	0.000000001 or 10^{-9}	$\dfrac{1\,nx}{10^{-9}\,x} = \dfrac{10^{-9}\,x}{1\,nx}$
pico-	p	0.000000000001 or 10^{-12}	$\dfrac{1\,px}{10^{-12}\,x} = \dfrac{10^{-12}\,x}{1\,px}$

Sample Exercises
1. *How many mm are there in 3.45 km?*
   The correct answer is 3.45 x 10^6 mm

The table above indicates there are 1000 m in 1 km and that 1 mm = 0.001 m.

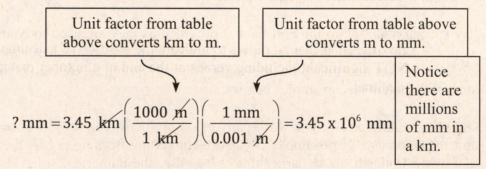

Unit factor from table above converts km to m.

Unit factor from table above converts m to mm.

Notice there are millions of mm in a km.

$$? \text{ mm} = 3.45 \text{ km} \left(\frac{1000 \text{ m}}{1 \text{ km}} \right) \left(\frac{1 \text{ mm}}{0.001 \text{ m}} \right) = 3.45 \times 10^6 \text{ mm}$$

Note how the km and m cancel. Unit canceling is one key to dimensional analysis problems.

The km unit is much larger than mm. We expect that there will be many millimeters (mm) in the large unit km unit. The answer, 3.45 x 10^6 mm, is sensible.

6

TIP   Always convert to a base unit (like m or g) first. Then proceed to different units as necessary.

2. How many mg are there in 15.0 pg?
   The correct answer is 1.5×10^{-8} mg

From the table we see that $1 \text{ pg} = 10^{-12}$ g and $1 \text{ mg} = 10^{-3}$ g.

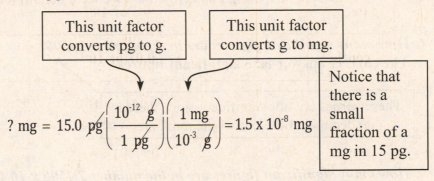

This unit factor converts pg to g.

This unit factor converts g to mg.

$$? \text{ mg} = 15.0 \text{ pg} \left(\frac{10^{-12} \text{ g}}{1 \text{ pg}} \right) \left(\frac{1 \text{ mg}}{10^{-3} \text{ g}} \right) = 1.5 \times 10^{-8} \text{ mg}$$

Notice that there is a small fraction of a mg in 15 pg.

In this problem pictogram (pg) is the smaller unit. We expect there to be very few milligrams (mg) in 15.0 pg. The answer of 1.5×10^{-8} mg is reasonable.

Significant Figures

All non-zero integers are significant.   When determining the number of significant figures in a value, by far the most confusion involves zeros. Sometimes zeroes are significant, and sometimes they are not! Below are the rules that will help you determine whether or not a zero is significant.

1. Zeros located between two integers **ARE** significant.
2. Zeros located at the ends of numbers containing decimals **ARE** significant.
3. Zeros located between an integer to the right and a decimal to the left **ARE NOT** significant.
4. Zeros used as place-holders to indicate the position of a decimal **ARE NOT** significant including zeroes at the end of a number that does not have a decimal.

Sample Exercises

3. How many significant figures are in the number 58062?
   The correct answer is: 5 significant figures

This zero is significant because it is embedded in other significant digits. See rule 1.

4. How many significant figures are in the number 0.0000543?
   The correct answer is: 3 significant figures

None of the zeroes in this number are significant because their purpose is to indicate the position of the decimal place. Only the non-zero integers 543 are significant.

7

5. How many significant figures are in the number 0.009120?
    The correct answer is: 4 significant figures

These three zeroes are not significant because they are place-holders.

TIPS

Zeroes at the end of a number including a decimal point are significant. The best way to remember this is to consider that we would not go to the trouble of writing that zero unless we were certain it had meaning!

6. How many significant figures are in the number 24500?
    The correct answer is: 3 significant figures

These zeroes are not significant since there is no decimal at the end of the number.

7. How many significant figures are in the number 2.4500×10^4?
    The correct answer is: 5 significant figures

As written both of these zeroes are significant because the number contains a decimal.

Notice that this number is the same as a previous exercise but written in scientific notation. All other rules apply.

CAUTION

For numbers written in scientific notation none of the numbers in the 10^x portion are significant.

Calculations and Significant Figures

Rules for determining the number of significant figures in the answer to a calculation depend on the mathematical operation being performed.

- In addition and subtraction problems, the final answer must contain no digits beyond the most doubtful digit in the numbers being added or subtracted.
- In multiplication and division problems involving significant figures the final answer must contain the same number of significant figures as the number with the least number of significant figures.

Sample Exercises

8. What is the sum of 12.674 + 5.3150 + 486.9?
    The correct answer is: 504.9

This 9 is in the tenths decimal place. It is the most doubtful digit in the sum.

8

The most doubtful digit in each of the numbers is underlined 12.674, 5.3150, 486.9. Notice that 486.9 has the most doubtful digit because the 9 is in the tenths position and the other numbers are doubtful in the thousandths and ten thousandths positions. *The final answer must have the final digit in the tenths position.*

9. *What is the correct answer to this problem: $2.6138 \times 10^6 - 7.95 \times 10^{-3}$?*
   The correct answer is: 2.6138×10^6

> This 8 is the most doubtful digit in the sum. It is in the hundreds position.

The number 2.6138×10^6 can be also written as 2,613,800. Its most doubtful digit, the 8, is in the hundreds position. The other number, 7.95×10^{-3}, can be written as 0.00795. Its most doubtful digit, the 5, is in the one millionths position. Consequently, the final answer cannot extend beyond the 8 in 2.6138×10^6.

When adding and subtracting, expressing both numbers in the same power of 10 helps determine the most doubtful digit.

10. *What is the correct answer to this problem: 47.893×2.64?*
   The correct answer is: 126

> This number contains only 3 significant figures, so the final answer can have only 3 significant figures.

11. *What is the correct answer to this problem: $1.95 \times 10^5 \div 7.643 \times 10^{-4}$?*
   The correct answer is: 2.55×10^8

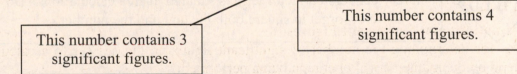

This number contains 3 significant figures.

This number contains 4 significant figures.

As in exercise 10, the number with fewest significant digits determines that the final answer has three significant digits.

Dimensional Analysis

In chemistry calculations frequently require changing from one set of units, (ft or in^2 or cm^3, for example) to a second set of units (Mm or km^2 or yd^3). Dimensional analysis is a convenient method to help convert units without making arithmetic errors. In this method, common conversion factors given in your textbook are arranged so that units cancel converting to a second set of units.

Sample Exercises

12. How many Mm are in 653 ft?

The correct answer is: 1.99 x 10⁻⁴ Mm.

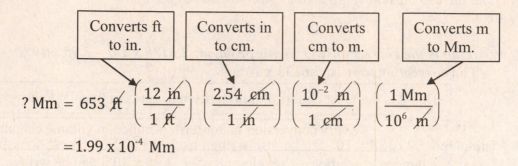

$$? \, Mm = 653 \, ft \left(\frac{12 \, in}{1 \, ft} \right) \left(\frac{2.54 \, cm}{1 \, in} \right) \left(\frac{10^{-2} \, m}{1 \, cm} \right) \left(\frac{1 \, Mm}{10^6 \, m} \right)$$

$$= 1.99 \times 10^{-4} \, Mm$$

Boxes label: Converts ft to in. | Converts in to cm. | Converts cm to m. | Converts m to Mm.

Notice the problem is arranged so that each successive conversion factor moves us further along in the conversion process. Feet are converted to inches, then to cm, next to m, and finally to Mm. This is an example of the simplest kind of dimensional analysis problem because all units are linear.

13. How many km² are in 2.5 x 10⁸ in²?

The correct answer is: 1.6 x 10⁻¹ km²

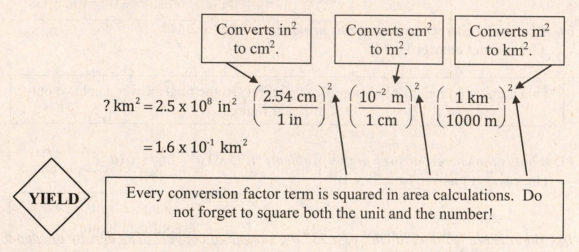

$$? \, km^2 = 2.5 \times 10^8 \, in^2 \left(\frac{2.54 \, cm}{1 \, in} \right)^2 \left(\frac{10^{-2} \, m}{1 \, cm} \right)^2 \left(\frac{1 \, km}{1000 \, m} \right)^2$$

$$= 1.6 \times 10^{-1} \, km^2$$

Boxes label: Converts in² to cm². | Converts cm² to m². | Converts m² to km².

YIELD — Every conversion factor term is squared in area calculations. Do not forget to square both the unit and the number!

Because this problem involves the two dimensional unit area all conversion factors are similar to exercise 12, but they are squared to have the appropriate units.

TIP — The ( )² notation around a unit factor literally means you are multiplying the unit factor by itself. If you have trouble remembering to square the unit factor, then try writing it as the unit factor multiplied by the unit factor.

14. How many yd³ are in 7.93 x 10¹² cm³?
   The correct answer is: 1.04 x 10⁷ yd³

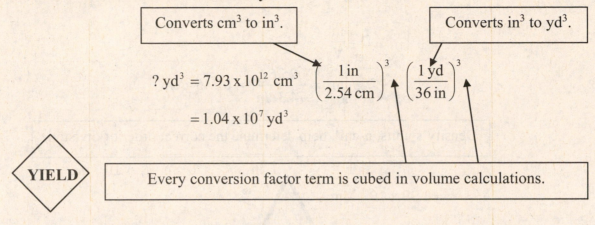

Converts cm³ to in³.

Converts in³ to yd³.

$$? \, yd^3 = 7.93 \times 10^{12} \, cm^3 \left(\frac{1 \, in}{2.54 \, cm}\right)^3 \left(\frac{1 \, yd}{36 \, in}\right)^3$$

$$= 1.04 \times 10^7 \, yd^3$$

YIELD

Every conversion factor term is cubed in volume calculations.

Density

15. What is the mass, in g, of a 68.2 cm³ sample of ethyl alcohol? The density of ethyl alcohol is 0.789 g/cm³.
   The correct answer is: 53.8 g

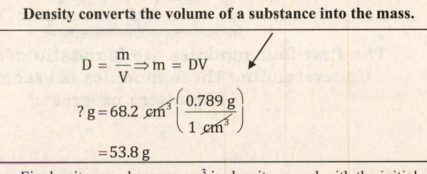

Density converts the volume of a substance into the mass.

$$D = \frac{m}{V} \Rightarrow m = DV$$

$$? \, g = 68.2 \, cm^3 \left(\frac{0.789 \, g}{1 \, cm^3}\right)$$

$$= 53.8 \, g$$

Final units are g because cm³ in density cancel with the initial volume units.

16. What is the volume, in cm³, of a 237.0 g sample of copper? The density of copper is 8.92 g/cm³.
   The correct answer is: 26.6. cm³

$$D = \frac{m}{V} \Rightarrow V = \frac{m}{D}$$

$$? \, cm^3 = 237.0 \, g \left(\frac{1 \, cm^3}{8.92 \, g}\right)$$

$$= 26.6 \, cm^3$$

17. What is the density of a substance with a mass of 25.6 g and volume of 74.3 cm³?
   The correct answer is: 0.345 g/cm³

$$D = \frac{m}{V}$$

$$? \, g/cm^3 = \frac{25.6 \, g}{74.3 \, cm^3} = 0.345 \, g/cm^3$$

Density's units, g/cm³, help determine the correct order of division.

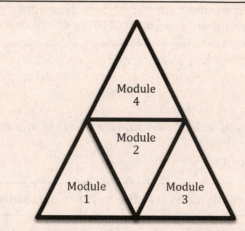

The first four modules are foundations of chemistry. Understanding these modules is essential to your chemistry progress.

Study Tip #2

Cognitive scientists know a great deal about how people learn. Here are some simple strategies to improve your learning.

1. Active learning is more lasting than passive learning
   Use techniques that engage your brain as you study. Look at Study Tip #4 for several active techniques.
2. Thinking about thinking is important
   Metacognition is the process of continually asking, "Do I understand this? What is my level of understanding? etc." See Study Tip #3.
3. Group work and talking are essential
   True learning occurs when your neurons fire as you think about chemistry. Form a study group and be an active group participant. In forming words to talk or ask a question your neurons fire stimulating learning.
4. The level at which learning occurs is important
   Bloom's Taxonomy, shown below, is one method to assess your learning level. Most college chemistry classes want you to take what was presented in class then apply it to new situations. Ask yourself if you can do that. It is a necessary skill to transition from high school to college.

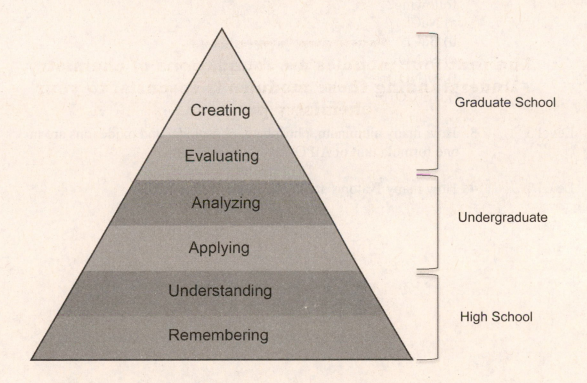

R. Hoffmann and S. Y. McGuire, "Learning and Teaching Strategies," American Scientist, Volume 98 (378-382), 2010.

Module 2 Predictor Questions

The following questions may help you determine the extent you need to study this module. Questions are ranked according to ability.

Level 1 = basic proficiency
Level 2 = mid level proficiency
Level 3 = high proficiency

If you can correctly answer Level 3 questions you probably do not need to spend much time on this module. If you can only answer Level 1 problems, you should review this module.

Level 1   1. How many atoms of each element are in one molecule or formula unit of each of the following?
   a) C_4H_9OH
   b) $MgBr_2$
   c) $Ba_3(PO_4)_2$

Level 1   2. How many ions are present in one formula unit of each of the following?
   a) $NaCl$
   b) $BaCl_2$
   c) $NaNO_3$
   d) $Al(NO_3)_3$
   e) $Al_2(CO_3)_3$

Level 3   3. How many aluminum, phosphate, phosphide, and oxide ions are in one formula unit of $AlPO_4$?

Level 1   4. How many P atoms are in one mole of $Mg_3(PO_4)_2$?

Module 2 Predictor Question Solutions

1. How many atoms of each element are in one molecule or formula unit of each of the following?
   a) C_4H_9OH
   b) $MgBr_2$
   c) $Ba_3(PO_4)_2$

   a) **One molecule of C_4H_9OH contains <u>four</u> C atoms, <u>ten</u> (9 +1) H atoms, and <u>one</u> O atom.**
   b) **One formula unit of $MgBr_2$ contains <u>one</u> Mg atom (ion) and <u>two</u> Br atoms (ions).**
   c) **One formula unit of $Ba_3(PO_4)_2$ contains <u>three</u> Ba atoms (ions), <u>two</u> P atoms (in the phosphate ions), and <u>eight</u> O atoms (in the phosphate ions).**

2. How many total ions are present in one formula unit of each of the following?
   a) NaCl
   b) $BaCl_2$
   c) $NaNO_3$
   d) $Al(NO_3)_3$
   e) $Al_2(CO_3)_3$

   a) **<u>Two</u> total ions (one Na^+ and one Cl^-) in one formula unit of NaCl.**
   b) **<u>Three</u> total ions (one Ba^{2+} and two Cl^-) in one formula unit of $BaCl_2$.**
   c) **<u>Two</u> total ions (one Na^+ and one NO_3^-) in one formula unit of $NaNO_3$.**
   d) **<u>Four</u> total ions (one Al^{3+} and three NO_3^-) in one formula unit of $Al(NO_3)_3$.**
   e) **<u>Five</u> total ions (two Al^{3+} and three CO_3^{2-}) in one formula unit of $Al_2(CO_3)_3$.**

3. How many aluminum, phosphate, phosphide, and oxide ions are present in one formula unit of $AlPO_4$?

   There is <u>one</u> aluminum (Al^{3+}) ion, <u>one</u> phosphate (PO_4^{3-}) ion, <u>zero</u> phosphide (P^{3-}), and zero oxide (O^{2-}) ions in one formula unit of $AlPO_4$. The phosphate ion is a polyatomic ion that does not further break down into ions or atoms, so there are no phosphide or oxide ions present.

4. How many P atoms are in one mole of $Mg_3(PO_4)_2$?

a) In one mole of $Mg_3(PO_4)_2$ there are two moles of P atoms. The number of atoms is

$$2 \, moles \, P \, atoms \left(\frac{6.022 \times 10^{23} \, P \, atoms}{1 \, mole \, P \, atoms} \right) = 1.2044 \times 10^{24} \, P \, atoms$$

Module 2
Understanding Chemical Formulas

Introduction

What information is contained in a chemical formula and how do we interpret that information? Chemists use specific symbolism to express their understanding of elements, compounds, ions, and ionic compounds. The primary goal of this module is to help you:

1. recognize these symbols
2. learn how to determine the number and types of atoms or ions present in a substance.

Module 2 Key Equations & Concepts

1. *Molecular formulas*
   Indicate the number of each **atom** present in a **molecule** (C_5H_{12}).
2. *Ionic formulas*
   Indicate the number of each **ion** present in a **formula unit** ($Al_2(CO_3)_3$)
   Ionic formulas also indicate the numbers of each element present in the formula unit.
3. *Stoichiometric coefficients*
   Indicate the amount of a particular molecule or formula unit in the chemical symbolism. Stoichiometric coefficients are used in balanced chemical equations.

Sample Exercises
Interpreting Chemical Formulas

1. How many atoms of each element are present in one molecule of C_2H_5OH?
   The correct answer is: 2 C, 6 H, and 1 O

$$C_2H_5OH$$

There are 2 carbon atoms, 1 oxygen atom, and 6 hydrogen atoms in one molecule of C_2H_5OH. Molecular formulas indicate numbers of atoms of each element.

TIP

Do not forget that if there is no subscript written it is understood there is one atom of that element present.

2. How many atoms of each element are present in one formula unit of $Al_2(SO_4)_3$?
   The correct answer is: 2 Al, 3 S, and 12 O

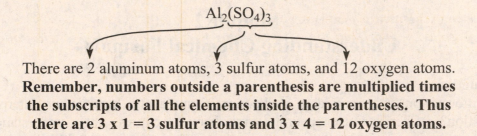

$$Al_2(SO_4)_3$$

There are 2 aluminum atoms, 3 sulfur atoms, and 12 oxygen atoms.
**Remember, numbers outside a parenthesis are multiplied times
the subscripts of all the elements inside the parentheses. Thus
there are 3 x 1 = 3 sulfur atoms and 3 x 4 = 12 oxygen atoms.**

This example is an *ionic compound* (see Module 3). The parentheses surrounding (SO_4) indicate it is a *polyatomic ion*. Its actual formula is SO_4^{2-}. Two Al^{3+} ions are required to balance the charge of the three SO_4^{2-}. So, this formula indicates there are 2 Al^{3+} ions and 3 SO_4^{2-} ions for a total of 5 ions.

Interpreting Stoichiometric Coefficients
3. *How many atoms of each element are present in 3 molecules of C_5H_{12}?*
   The correct answer is: 15 C, 36 H

$$3\ C_5H_{12}$$

There are 15 carbon atoms and 36 hydrogen atoms in 3 C_5H_{12} molecules.
3 x 5 = 15 C atoms   3 x 12 = 36 H atoms

The 3 is a *stoichiometric coefficient* such as used in balanced chemical equations.

4. *How many atoms of each element are present in five formula units of $Ca_3(PO_4)_2$?*
   How many ions are in one formula unit of $Ca_3(PO_4)_2$?
   **The correct answer is: 15 Ca, 10 P, and 40 O in five formula units; 5 ions in one
   formula unit.**

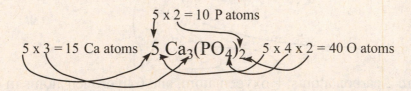

5 x 2 = 10 P atoms

5 x 3 = 15 Ca atoms $5\ Ca_3(PO_4)_2$ 5 x 4 x 2 = 40 O atoms

There are 15 Ca atoms, 10 P atoms, and 40 O atoms in 5 formula units of $Ca_3(PO_4)_2$.
The ionic formula also indicates there are 3 Ca^{2+} ions and 2 PO_4^{3-} ions in one formula
unit of $Ca_3(PO_4)_2$.

⚠ TIP   Remember the stoichiometric coefficient is multiplied by *every* subscript.

18

Interpreting Chemical Formulas

5. *Using circles to represent atoms draw your best representation of what one C_4H_{10} molecule looks like if we could see atoms, ions, and molecules.*

C_4H_{10}

The 4 carbon atoms are in the center of the molecule.

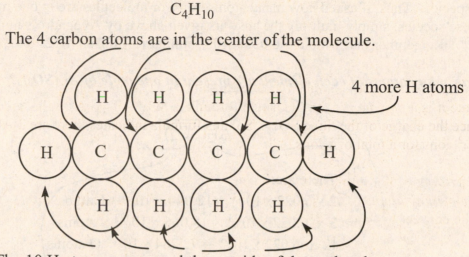

4 more H atoms

The 10 H atoms are around the outside of the molecule.
Notice, that this is one single molecule not 14 separate things.

From a chemical standpoint this is not the only way to draw C_4H_{10}, but all of the possibilities consist of molecules with bonded (connected) atoms.

6. *Using circles to represent atoms and ions draw your best representation of what $Sr_3(PO_4)_2$ looks like if we could see atoms, ions, and molecules. Remember, ions are independent species.*

Notice that the three Sr^{2+} ions are independent species.

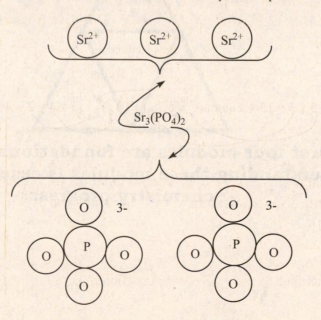

The two PO_4^{3-} ions are also independent species.

19

Using Chemical Formulas to Determine Atom Numbers in One Mole of a Substance

All of the formulas and symbols introduced up to now are also used to represent moles of a species. Thus, if asked how many atoms, ions, or molecules are in one mole of each of these species, simply multiply the answers given above by Avogadro's number, 6.022 x 10^{23}. Remember that 1 mole = 6.022 x 10^{23}, similar to 1 dozen = 12.

7. How many atoms of each element are present in one mole of $Al_2(SO_4)_3$?

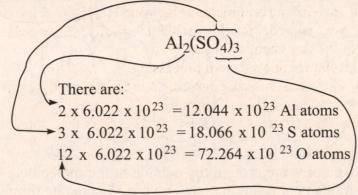

$$Al_2(SO_4)_3$$

There are:

2 x 6.022 x 10^{23} = 12.044 x 10^{23} Al atoms

3 x 6.022 x 10^{23} = 18.066 x 10^{23} S atoms

12 x 6.022 x 10^{23} = 72.264 x 10^{23} O atoms

Chemical formulas contain lots of information. You must understand the difference between atoms, ions, and molecules to correctly understand how to apply all that information.

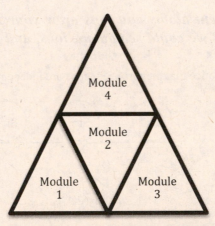

Module 4

Module 2

Module 1   Module 3

The first four modules are foundations of chemistry. Understanding these modules is essential to your chemistry progress.

Study Tip #3

Metacognition is the ability to:

1. Think about one's own thinking

   As you are studying ask yourself, "Am I understanding this correctly? Does this knowledge connect to other things I know about chemistry? How will we use this later in the course?"

2. Be consciously aware of oneself as a problem solver

   As you are working on problems ask yourself, "Does this answer make sense? Am I approaching this problem logically? How could I apply this problem to other chemical topics?"

3. Monitor and control one's mental processing

   During study sessions ask yourself, "Am I understanding this material?"

4. Accurately judge one's level of learning

   Can you take an idea or topics presented in class then apply them to a new problem?

J. H. Flavell, Metacognitive aspects of problem solving. ...
Alberta Journal of Educational Research, 26 (169-182), 1976.

Module 3 Predictor Questions

The following questions may help you determine the extent you need to study this module. Questions are ranked according to ability.

      Level 1 = basic proficiency

      Level 2 = mid level proficiency

      Level 3 = high proficiency

If you can correctly answer Level 3 questions you probably do not need to spend much time on this module. If you can only answer Level 1 problems, you should review this module.

1. Name the following inorganic compounds.

      Level 1   a) N_2O_5

      Level 1   b) $MgCl_2$

      Level 2   c) $Al(OH)_3$

      Level 2   d) $HClO_4$

      Level 3   e) $NaClO_3$

      Level 3   f) $KHSO_4$

2. Write the formulas of the following inorganic compounds.

      Level 1   a) sulfur trioxide

      Level 1   b) cesium bromide

      Level 2   c) ammonium chloride

      Level 2   d) nitric acid

      Level 3   e) potassium sulfate

      Level 3   f) calcium hydrogen phosphate

3. Determine the number of ions in one formula unit of each of the following compounds.

      Level 1   a) CO_2

      Level 1   b) H_3PO_4

      Level 2   c) $Al(OH)_3$

      Level 3   d) KNO_2

      Level 3   e) $Al(OH)Cl_2$

Module 3 Predictor Question Solutions

1. Name these inorganic compounds.

Compound	Answer
a) N_2O_5	**dinitrogen pentoxide**
b) $MgCl_2$	**magnesium chloride**
c) $Al(OH)_3$	**aluminum hydroxide**
d) $HClO_4$	**perchloric acid**
e) $NaClO_3$	**sodium chlorate**
f) $KHSO_4$	**potassium hydrogen sulfate**

2. Write chemical formulas of these inorganic compounds.

Compound	Answer
a) sulfur trioxide	**SO_3**
b) cesium bromide	**$CsBr$**
c) ammonium chloride	**NH_4Cl**
d) nitric acid	**HNO_3**
e) potassium sulfate	**K_2SO_4**
f) calcium hydrogen phosphate	**$CaHPO_4$**

3. Determine the number of ions in one molecule or formula unit of each of these compounds.

Compound	Answer
a) CO_2	**0 CO_2 is a covalent compound; there are no ions present**
b) H_3PO_4	**0 (when not dissolved in water); when dissolved in water H_3PO_4 dissociates into 3 H^+ and 1 PO_4^{3-}; there are 4 ions**
c) $Al(OH)_3$	**4 (one Al^{3+} and three OH^-)**
d) KNO_2	**2 (one K^+ and one NO_2^-)**
e) $Al(OH)Cl_2$	**4 (one Al^{3+}, one OH^-, and two Cl^-)**

Module 3
Chemical Nomenclature

Introduction

Chemical nomenclature is a combination of chemical symbolic and naming languages. To ensure that each chemical name is interpreted correctly, chemists follow a specific set of nomenclature rules. This module will:

1. familiarize you with the rules of chemical nomenclature
2. help you recognize various types of chemical compounds then apply the appropriate nomenclature rules.

Module 3 Key Concepts

Ionic compounds:

Ionic compounds are composed of a metal (or ammonium) cation and one or more anions. There are several types of ionic compounds, each with their own rules for naming.

Cations combined with:

1. **Nonmetal anions** (simple binary ionic compounds)
   Nomenclature is the metal's name followed by nonmetal stem plus –ide.
   If the metal cation is a transition metal add the oxidation state in parentheses after the metal's name.
2. **Polyatomic anions** (pseudobinary ionic compounds)
   Nomenclature is cation name followed by polyatomic ion name.

 TIP: Consult your textbook for a list of common polyatomic ions whose names and formulas you should recognize.

Covalent compounds

Covalent compounds are composed of two or more nonmetals.

3. **Two nonmetals** (binary covalent compounds)
   The less electronegative element is named first, and the more electronegative is named second using stem plus –ide. Prefixes such as di-, tri-, etc. are used for both elements.
4. **Hydrogen combined with a nonmetal in aqueous solution** (binary acid)
   Nomenclature is hydro followed by nonmetal stem with suffix –ic acid.
5. **Hydrogen, oxygen, and a nonmetal combined in one compound** (ternary acids)
   Nomenclature is a series of names based upon the oxidation state of the nonmetal.
   Nonmetal highest oxidation state is **per** *stem* **–ic acid**.
   Nonmetal second highest oxidation state is *stem* **–ic acid**.
   Nonmetal third highest oxidation state is *stem* **–ous acid**.
   Nonmetal lowest oxidation state is **hypo** *stem* **–ous acid**.

Special ionic compound types

6. **Metal ions combined with a polyatomic ion made from a ternary acid**
   Ternary acid salts – nomenclature is the metal name followed by the same series of names used for the ternary acid with two changes.
   –ic suffixes are changed to –ate
   –ous suffixes are changed to –ite
7. **Metal ions combined with a ternary acid salt and hydrogen**.
   Acidic salts of ternary acids – nomenclature is the metal name followed by hydrogen (including the appropriate di-, tri-, etc. prefix) plus the ternary acid salt name.
8. **Metal ions combined with hydroxyl groups and nonmetal ions**.
   Basic salts of polyhydroxy bases – nomenclature is the metal name followed by hydroxy (including the appropriate di-, tri-, etc. prefix) plus the nonmetal stem plus –ide.

Sample Exercises

1. *What is the correct name of the chemical compound CaBr₂?*
   The correct answer is: calcium bromide

Metal cations and nonmetal anions form <u>simple binary ionic compounds</u>. Simple binary ionic compounds are named using the metal name followed by the nonmetal stem with the suffix –ide. *Prefixes like di- or tri- are **not** used to denote ion numbers present in the substance*.

Elemental Stems for Common Monatomic Anions

				H hydr
B bor-	C carb-	N nitr-	O ox-	F fluor-
	Si silic-	P phosph-	S sulf-	Cl chlor-
		As arsen-	Se selen-	Br brom-
		Sb antimon-	Te tellur-	I iod-

The metal cation in this case is Ca^{2+}, calcium. The anion, Br^-, is named using the brom stem plus –ide.

2. *What is the correct name of the chemical compound Mg₃(PO₄)₂?*
   The correct answer is: magnesium phosphate

Metal cations and polyatomic anions form <u>pseudobinary ionic compounds</u>. These compounds are named using the metal name followed by the correct name of the polyatomic anion. Your textbook has a list of polyatomic anions you must know. Make sure you know the name, formula, and charge of the anion. *Once again, no prefixes are used to indicate ion numbers in these compounds*.

Mg^{2+} is a positive ion made from the metal magnesium. PO_4^{3-} is a negative polyatomic ion named phosphate.

3. *What is the correct name of this chemical compound, $FeCl_3$?*
   The correct answer is: iron (III) chloride

Transition metal cations and nonmetal or polyatomic anions form <u>transition metal ionic compounds</u>. Their names are derived from the metal name followed by the metallic oxidation state using Roman numerals inside parentheses. The metal oxidation state is determined using the anion oxidation state.
Fe^{3+} is a positive ion made from a transition metal. Cl^- is a negative ion made from the nonmetal chlorine.

4. *What is the correct name of the chemical compound, N_2O_4?*
   The correct answer is: dinitrogen tetroxide
This compound is made from two nonmetals, nitrogen and oxygen, so it is a <u>binary covalent compound</u>. *Binary covalent compounds use prefixes to indicate the number of atoms of each element present in the compound.* This is an important difference from the previous ionic compound examples.

5. *What is the correct name of the chemical compound $H_2S(aq)$?*
   The correct answer is: hydrosulfuric acid

This compound is made from hydrogen and a nonmetal. Furthermore, the (aq) symbol indicates this compound is dissolved in water. That combination indicates a <u>binary acid</u>. Binary acids are named using the prefix hydro- followed by the nonmetal's stem and the suffix –ide.

If the symbol (aq) is not present the compound is named as a binary covalent compound. In this case, without the (aq), H_2S is named dihydrogen sulfide.

6. *What is the correct name of this chemical compound, $HClO_3$?*
   The correct answer is: chloric acid

This compound is made from three nonmetals, H, O, and another nonmetal, chlorine. This combination of nonmetals indicates a <u>ternary acid</u>. Ternary acid names are based upon the oxidation state of the nonmetal other than H and O. The easiest method to learn these compounds is to use the following table of "-ic acids". You must learn both the compound formula and its name.

Names and Formulas of the Common Ternary –ic Acids

IIIA	IVA	VA	VIA	VIIA
H_3BO_3 boric acid	H_2CO_3 carbonic acid	HNO_3 nitric acid		
	H_2SiO_3 silicic acid	H_3PO_4 phosphoric acid	H_2SO_4 sulfuric acid	$HClO_3$ chloric acid
		H_3AsO_4 arsenic acid	H_2SeO_4 selenic acid	$HBrO_3$ bromic acid
			H_6TeO_6 telluric acid	HIO_3 iodic acid

Once you know the "-ic acids" use the following system:
The acid with one more O atom than the "-ic acid" is the "**per** *stem* **-ic acid**".
One fewer O atom than the "-ic acid" is the "*stem* **-ous acid**".
Two fewer O atoms than the "-ic acid" is the "**hypo** *stem* **-ous acid**".

The chlorine acid series is given below:
$HClO_4$ is perchloric acid.
$HClO_3$ is chloric acid.
$HClO_2$ is chlorous acid.
$HClO$ is hypochlorous acid.

7. *What is the correct name of the chemical compound $KClO_4$?*
   The correct answer is: potassium perchlorate

This compound is made from a metal ion, K^+, and a polyatomic anion derived from one of the ternary acids discussed above. $KClO_4$ is a <u>ternary acid salt</u>. The anion name is based upon the ending of the ternary acid. Ternary acids ending in "-ic" form salts ending with "ate". Ternary acids ending with "ous" form salts endin in "ite". The prefixes per- and hypo- are retained.

The potassium salts of the chlorine acid series are given below:
$KClO_4$ is potassium perchlorate.
$KClO_3$ is potassium chlorate.
$KClO_2$ is potassium chlorite.
$KClO$ is potassium hypochlorite.

CAUTION   These compounds are some of the most difficult to name. Pay special attention to them.

8. What is the correct name of the chemical compound NaH₂PO4?
   The correct answer is: sodium dihydrogen phosphate

This compound is made from a metal cation, Na^+, and a polyatomic anion made from a ternary acid retaining some acidic hydrogens. Such compounds are <u>acidic salts of ternary acids</u>. Nomenclature for these compounds uses the word hydrogen plus a prefix, in this case di-, to indicate the number of acidic hydrogens present. The last part of the salt name is the same determined in exercise 7 for ternary acid salts.

9. What is the correct name of the chemical compound Al(OH)₂Cl?
   The correct answer is: aluminum dihydroxy chloride

This compound is made from a metal ion, Al^{3+}, and three anions (two hydroxide and one chloride ion). Compounds containing hydroxide ions and other anions plus a metal ion are <u>basic salts of polyhydroxy bases</u>. Their name indicates the number of OH^- groups present in the compound. This is done using the appropriate prefix attached to hydroxy. The rest of the compound name is the same as for binary ionic compounds.

1. If the compound contains a metal cation, then you should NOT use prefixes in the name. Prefixes are used only in covalent compounds.
2. Check the location of the metal on the periodic table. If the metal is a transition metal, then you likely need to use a Roman numeral to indicate its oxidation state (there are a few exceptions; see your textbook).
3. Be sure you are familiar with the names and formulas of the common polyatomic ions.
4. Pay careful attention to whether a binary acid is written with (g) or (aq). Binary acids have different names depending on those symbols!
5. Be sure you know the names and formulas (including charges) of the "-ic acids and the "-ate ions."

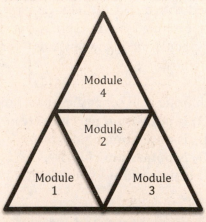

**The first four modules are foundations of chemistry.
Understanding these modules is essential to your
chemistry progress.**

Study Tip #4

Adjust your study methods for college success
1. Three very common poor study methods are:
    Highlighting using a colored pen.
    Read and reread the chapter.
    Make a set of notecards.
2. Much better study techniques are:
    Annotate your textbook in the margins. Take the author's words, convert them into your own words, and write it in the margin. Use as few words as possible. When preparing for a test this will be your "notecards."
    Make concept maps showing how the topics in a textbook are related to one another. A concept map based upon a section of Chapter 2 in Kotz, Treichel, and Townsend, Seventh Edition is shown below.
    Work problems in the back of each chapter then design your own problems. Ask yourself how your instructor could change this problem to a new one. For example in a density problem you are given the mass and volume. If your instructor asks a problem giving the density and mass, what would it look like?
    Form a study group and be an active participant.

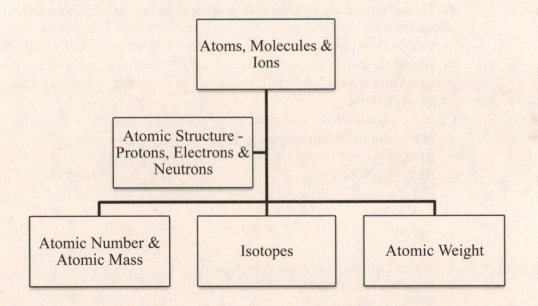

Practice Test One
Modules 1-3

Level 2   1. How many dm^3 are present in 3.24×10^8 in^3?

Level 2   2. Perform this numerical calculation to the correct number of significant figures. Determine how many significant figures are in each number.

$$((27.340 - 6.00) \times (6.8371 \times 10^3)) + 871.4$$

Level 1   3. The density of mercury is 13.59 g/cm^3. What is the mass of a mercury sample with a volume = 0.0230 in^3?

Level 1   4. How many nitrogen, N, atoms are present in 2.57 mol of $Al(NO_3)_3$?

Level 1   5. How many moles of $Ca_3(AsO_4)_2$ are present in 723.2 g of $Ca_3(AsO_4)_2$?

Level 3   6. How many of each of the following species are in one formula unit of $Fe_2(SO_4)_3$?
   a) iron (III) ions
   b) sulfate ions
   c) sulfide ions
   d) oxide ions

Levels 1-3   7. Name the following compounds.
   a) PCl_5
   b) $(NH_4)_2SO_4$
   c) $LiNO_3$
   d) KH_2BO_2
   e) XeF_4

Levels 1-3   8. Write formulas for these compounds.
   a) sulfur hexafluoride
   b) hydrocyanic acid
   c) copper (II) monohydroxy chloride
   d) magnesium bromide
   e) hypochlorous acid

Module 4 Predictor Questions

Level 1   1. Determine the formula weight of $Mg(ClO_3)_2$.

Level 2   2. How many molecules of C_2H_5OH are in 0.342 mols of C_2H_5OH?

Level 3   3. How many H atoms are in 0.342 mols of C_2H_5OH?

Level 3   4. How many nitrate ions are in 0.147 mols of $Zn(NO_3)_2$?

Level 1   5. What is the mass, in grams, of 0.348 mols of C_3H_7OH?

Level 1   6. What is the mass, in grams, of 6.34×10^{22} atoms of Sc?

Level 3   7. What is the total mass, in grams, of the Cl atoms in 0.483 mols of $Ba(ClO_4)_2$?

Level 2   8. Determine the molar mass of the compound $Al_2(SO_4)_3$.
                     a) How many mols of $Al_2(SO_4)_3$ are present in 94.2 g of $Al_2(SO_4)_3$?
                     b) How many O atoms are present in 37.5 g of $Al_2(SO_4)_3$.

31

Module 4 Predictor Question Solutions

1. Determine the molar mass of C_6H_{14}.

   For C atoms $6 \times 12.01\,g/mol = 72.06\,g$

   For H atoms $14 \times 1.008\,g/mol = 14.11\,g$

   Molar mass $= 72.06\,g + 14.11\,g = 86.17\,g/mol$

2. Determine the formula weight of $Mg(ClO_3)_2$.

   For Mg, 1 mole $= 24.30\,g$

   For Cl, 2 moles $= 2\,moles \times 35.45\,g/mol = 70.90\,g$

   For O, 6 moles $= 6\,moles \times 16.00\,g/mole = 96.00\,g$

   Formula weight $= 24.30\,g + 70.90\,g + 96.00\,g = 191.20\,g$

3. How many molecules of C_2H_5OH are in 0.342 mols of C_2H_5OH?

   In 1.000 mole of C_2H_5OH there are 6.022×10^{23} C_2H_5OH molecules.

   $$0.342\,mols\,of\,C_2H_5OH\left(\frac{6.022 \times 10^{23}\,C_2H_5OH\,molecules}{1.000\,moles\,of\,C_2H_5OH}\right) = 2.06 \times 10^{23}\,C_2H_5OH\,molecules$$

4. How many H atoms are in 0.342 mols of C_2H_5OH?

   $$2.06 \times 10^{23}\,C_2H_5OH\,molecules\left(\frac{6\,H\,atoms}{1\,C_2H_5OH\,molecule}\right) = 1.24 \times 10^{24}\,H\,atoms$$

5. How many nitrate ions are in 0.147 mols of $Zn(NO_3)_2$?

   $$0.147\,mols\,Zn(NO_3)_2\left(\frac{6.022 \times 10^{23}\,Zn(NO_3)_2\,formula\,units}{1\,mol\,Zn(NO_3)_2}\right) = 8.85 \times 10^{22}\,Zn(NO_3)_2\,formula\,units$$

   $$8.85 \times 10^{22}\,Zn(NO_3)_2\,formula\,units\left(\frac{2\,NO_3^-\,ions}{1\,Zn(NO_3)_2\,formula\,units}\right) = 1.77 \times 10^{23}\,NO_3^-\,ions$$

6. What is the mass, in grams, of 0.348 mols of C_3H_7OH?

$$\text{Molar mass of } C_3H_7OH = (3 \times 12.01\,g/mol) + (8 \times 1.008\,g/mol) + (1 \times 16.00\,g/mol) = 60.09\,g/mol$$

$$0.348\,mols(60.09\,g/mol) = 20.91\,g$$

7. What is the mass, in grams, of 6.34×10^{22} atoms of Sc?

$$6.34 \times 10^{22}\,\text{atoms Sc} \left(\frac{1\,mol\,Sc}{6.022 \times 10^{23}\,atoms} \right) = 0.105\,\text{moles Sc}$$

$$0.105\,\text{moles Sc} \left(\frac{44.96\,g\,Sc}{1\,mol\,Sc} \right) = 4.73\,g\,Sc$$

8. What is the total mass, in grams, of the Cl atoms in 0.483 mols of $Ba(ClO_4)_2$?

$$0.483\,\text{mols } Ba(ClO_4)_2 \left(\frac{2\,mols\,Cl\,atoms}{1\,mol\,Ba(ClO_4)_2} \right) = 0.966\,\text{mols Cl atoms}$$

$$0.966\,\text{mols Cl atoms} \left(\frac{35.45\,g\,Cl}{1\,mol\,Cl\,atoms} \right) = 34.2\,g\,Cl$$

9. Determine the molar mass of the compound $Al_2(SO_4)_3$.

$$\text{Molar Mass of } Al_2(SO_4)_3 = (2 \times 26.98\,g) + (3 \times 32.07\,g) + (12 \times 16.00) = 342.17\,g$$

a) How many mols of $Al_2(SO_4)_3$ are present in 94.2 g of $Al_2(SO_4)_3$?

$$94.2\,g\,Al_2(SO_4)_3 \left(\frac{1\,mol\,Al_2(SO_4)_3}{342.17\,g\,Al_2(SO_4)_3} \right) = 0.275\,\text{mols } Al_2(SO_4)_3$$

b) How many O atoms are present in 37.5 g of $Al_2(SO_4)_3$

$$37.5\,g\,Al_2(SO_4)_3 \left(\frac{1\,mol\,Al_2(SO_4)_3}{342.17\,g\,Al_2(SO_4)_3} \right) = 0.110\,\text{mols } Al_2(SO_4)_3$$

$$0.110\,\text{mols } Al_2(SO_4)_3 \left(\frac{12\,mols\,O\,atoms}{1\,mol\,Al_2(SO_4)_3} \right) = 1.32\,\text{mols O atoms}$$

$$1.32\,\text{mols O atoms} \left(\frac{6.022 \times 10^{23}\,O\,atoms}{1\,mol\,O\,atoms} \right) = 7.95 \times 10^{23}\,\text{O atoms}$$

Module 4
The Mole Concept

Introduction

In this module we will examine several equations used in mole concept problems. The goal of this module is to teach you:

1. how to interpret, use, and perform all the important calculations involving the mole.

You will need a periodic table as you work these exercises. Atomic weights are found on a periodic table.

Module 4 Key Equations & Concepts

1. **Molar mass** $= \sum$ **atomic weights of atoms in a compound, molecule, or ion**

   The molar mass, molecular weight, or formula weight[*] are calculated by summing the atomic weights of the elements in the compound. Molar mass is the mass in grams of one mole of a substance.

2. **One mole $= 6.022 \times 10^{23}$ particles**

   Avogadro's relationship converts from the number of moles of a substance to the number of atoms, ions, or molecules of that substance and vice versa.

3. **Mass of one atom of an element** $= \left(\dfrac{\text{mass of an element}}{\text{1 mole of an element}} \right)\left(\dfrac{\text{1 mole of atoms}}{6.022 \times 10^{23} \text{ atoms}} \right)$

   The mass of one atom, ion, or molecule is used to determine the mass of a few atoms, ions, or molecules of a substance. Notice that the fraction in the first set of parentheses simply is a representation of molar mass.

4. **Mole ratio**

   A compounds chemical formula is the *ratio* of the different types of atom in the compound. The mole ratio is used to convert from mass or moles of a compound to mass or moles of a specific atom in the compound.

---

*The terms molar mass, molecular weight, and formula weight all apply to the same concept/calculation. Technically, the term molecular weight should be used only with covalent compounds and formula weight applies only to ionic compounds. The more generic term *molar mass* is used frequently in chemical literature.

Sample Exercises
Determining Molar Mass

1. What is the molar mass (or formula weight) of calcium phosphate, $Ca_3(PO_4)_2$?
   The correct answer is: 310.2 g/mol

$$Ca_3(PO_4)_2$$

Molar mass of $Ca_3(PO_4)_2 = (3 \times 40.08 \text{ g/mol Ca}) + (4 \times 2 \times 16.00 \text{ g/mol O}) + 2 \times 30.97 \text{ g/mol P}$
$= 310.18 \text{ g /mol } Ca_3(PO_4)_2$

Determining Number of Moles

2. *How many moles of calcium phosphate are in 65.3 g of Ca₃(PO₄)₂?*
   The correct answer is: 0.211 mol Ca₃(PO₄)₂

$$? \text{ moles of } Ca_3(PO_4)_2 = 65.3 \text{ g } Ca_3(PO_4)_2 \left(\frac{1 \text{ mol } Ca_3(PO_4)_2}{310.18 \text{ g } Ca_3(PO_4)_2} \right)$$

$$= 0.211 \text{ mol } Ca_3(PO_4)_2$$

Molar mass of calcium phosphate from exercise #1.

From the number of moles of a sample, we can determine the number of molecules or formula units of the substance. (Molecules are found in covalent compounds. Ionic compounds do not have molecules. Their smallest subunits are formula units.)

Determining Number of Molecules or Formula Units

3. *How many formula units of calcium phosphate are in 0.211 moles of Ca₃(PO₄)₂?*

$$? \text{ formula units of } Ca_3(PO_4)_2 = 0.211 \text{ moles of } Ca_3(PO_4)_2 \left(\frac{6.022 \times 10^{23} \text{ formula units}}{1 \text{ mole of } Ca_3(PO_4)_2} \right)$$

$$= 1.27 \times 10^{23} \text{ formula units of } Ca_3(PO_4)_2$$

Avogadro's relationship

CAUTION

Be careful with the labels! You have just calculated the number of *formula units*. Do not confuse this with the number of atoms or the number of ions! All are valid questions with different answers!

Determining Number of Atoms or Ions

4. *How many oxygen, O, atoms are there in 0.211 moles of Ca₃(PO₄)₂?*
   The correct answer is: 1.02 x 10²⁴ oxygen atoms

Using the last idea from the key concepts box, we can determine the mass of a few molecules or formula units of a compound.

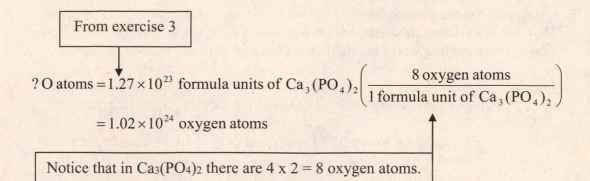

From exercise 3

$$? \text{ O atoms} = 1.27 \times 10^{23} \text{ formula units of } Ca_3(PO_4)_2 \left(\frac{8 \text{ oxygen atoms}}{1 \text{ formula unit of } Ca_3(PO_4)_2} \right)$$

$$= 1.02 \times 10^{24} \text{ oxygen atoms}$$

Notice that in $Ca_3(PO_4)_2$ there are 4 x 2 = 8 oxygen atoms.

Determining Mass of Molecules or Formula Units of a Substance

5. *What is the mass, in grams, of 25.0 formula units of $Ca_3(PO_4)_2$?*
   The correct answer is: 1.29 x 10⁻²⁰ g

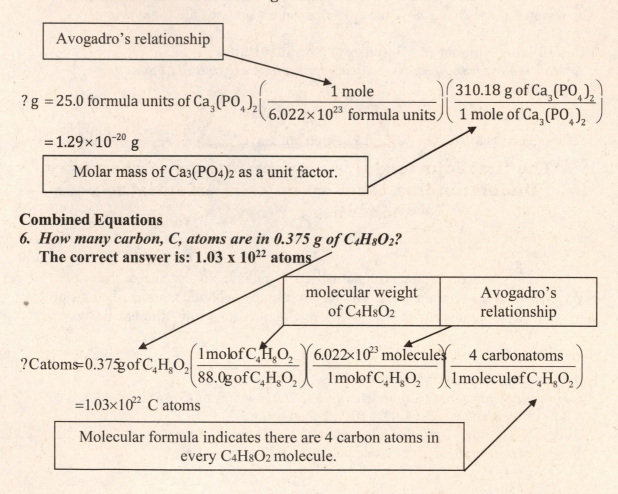

Avogadro's relationship

$$? \text{ g} = 25.0 \text{ formula units of } Ca_3(PO_4)_2 \left(\frac{1 \text{ mole}}{6.022 \times 10^{23} \text{ formula units}} \right) \left(\frac{310.18 \text{ g of } Ca_3(PO_4)_2}{1 \text{ mole of } Ca_3(PO_4)_2} \right)$$

$$= 1.29 \times 10^{-20} \text{ g}$$

Molar mass of $Ca_3(PO_4)_2$ as a unit factor.

Combined Equations

6. *How many carbon, C, atoms are in 0.375 g of $C_4H_8O_2$?*
   The correct answer is: 1.03 x 10²² atoms

molecular weight of $C_4H_8O_2$	Avogadro's relationship

$$? \text{ C atoms} = 0.375 \text{ g of } C_4H_8O_2 \left(\frac{1 \text{ mol of } C_4H_8O_2}{88.0 \text{ g of } C_4H_8O_2} \right) \left(\frac{6.022 \times 10^{23} \text{ molecules}}{1 \text{ mol of } C_4H_8O_2} \right) \left(\frac{4 \text{ carbon atoms}}{1 \text{ molecule of } C_4H_8O_2} \right)$$

$$= 1.03 \times 10^{22} \text{ C atoms}$$

Molecular formula indicates there are 4 carbon atoms in every $C_4H_8O_2$ molecule.

TIPS

A commonly encountered problem for students is deciding where to start on mole problems. If you have trouble getting started, focus on the given information. All of the examples in this module began by using the mass or number of moles stated in the question. You will frequently use some combination of molar masses, Avogadro's relationship, and mole ratio to solve mole problems. Select which entity to use first by looking at the unit given in the problem then determine, using dimensional analysis, how to cancel that unit.

Pay attention to vocabulary! Keep in mind the differences between atoms, ions, and molecules, and pay attention to which of the three the question is asking about.

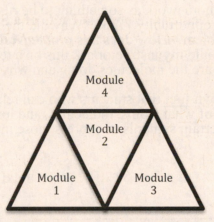

The first four modules are foundations of chemistry. Understanding these modules is essential to your chemistry progress.

Study Tip #5

1. Read your textbook daily

   Constantly refresh your memory about chemistry. When you sit down to study leave outside influences behind. Do not look at your cell phone, check Facebook, etc.

2. Do NOT work the practice test or homework over and over!

   This falsely convinces you that you understand the material when all you are doing is memorizing a set of steps. If your instructor changes the problem you will be lost.

3. DO rework homework problems that you missed the first time to understand why you missed it!

   Don't think of homework as something to be rushed through. Instead look at it as a learning opportunity. Ask, "Why did the instructor give me this homework problem? How does this problem relate to what we have done in class? How would my instructor ask this on a test?"

4. Ask yourself what are the molecules doing and why does this equation explain it?

   It is a well known **fact** that students who can relate classroom learning to mental images of what atoms, molecules, and ions look like and why chemists use certain symbols to express those images do very well in chemistry.

Module 5 Predictor Questions

The following questions may help you determine the extent you need to study this module. Questions are ranked according to ability.

      Level 1 = basic proficiency
      Level 2 = mid level proficiency
      Level 3 = high proficiency

If you can correctly answer Level 3 questions you probably do not need to spend much time on this module. If you can only answer Level 1 problems, you should review this module.

Level 1   1. Balance the following equations with the **smallest whole number coefficients**.

      a) ___$Fe(NO_3)_3$ + ___NH_3 + ___H_2O → ___$Fe(OH)_3$ + ___NH_4NO_3

      b) ___$C_2H_8N_2$ + ___N_2O_4 → ___N_2 + ___CO_2 + ___H_2O

Level 1   2. Given the balanced chemical reaction: $SiCl_4 + 2\,Mg \rightarrow Si + 2\,MgCl_2$, how many grams of Si could be produced by reacting 1.46 kg of $SiCl_4$ with excess Mg?

Level 1   3. If 58 moles of NH_3 are combined with 32 moles of sulfuric acid, what is the limiting reactant and how much of the excess reactant remains?
$$2\,NH_3 + H_2SO_4 \rightarrow (NH_4)_2SO_4$$

Level 1   4. What is the percent yield if 28.50 g of FeO reacts with excess CO and produces 17.841 g of Fe?
$$FeO + CO \rightarrow Fe + CO_2$$

Level 1   5. What volume of 0.158 M HBr solution is required to react completely with 38.77 mL of 0.226 M $Ca(OH)_2$ in the following reaction:
$$Ca(OH)_2 + 2HBr \rightarrow CaBr_2 + 2\,H_2O$$

Level 2   6. How many mL of 5.44 M $Sr(OH)_2$ are required to make 100.99 mL of a 0.189 M $Sr(OH)_2$ solution? What is the molar concentration of the Sr^{2+} ions in the 0.189 M solution? What is the molar concentration of the OH^- ions in the 0.189 M solution?

Module 5 Predictor Question Solutions

1. Balance the following equations with the **smallest whole number coefficients**.

   ___$Fe(NO_3)_3$ + **3** NH_3 + **3** H_2O → ___$Fe(OH)_3$ + **3** NH_4NO_3

   ___$C_2H_8N_2$ + ___N_2O_4 → **2** N_2 + **2** CO_2 + **4** H_2O

2. Given the balanced chemical reaction: $SiCl_4 + 2\ Mg → Si + 2\ MgCl_2$, ho many grams of Si could be produced by reacting 1.46 kg of $SiCl_4$ with excess Mg?

$$\textbf{1.46 kg SiCl}_4 \left(\frac{\textbf{1000 g SiCl}_4}{\textbf{1 kg SiCl}_4} \right) \left(\frac{\textbf{1 mol SiCl}_4}{\textbf{92.09 g SiCl}_4} \right) = \textbf{15.9 mol SiCl}_4$$

$$\textbf{15.9 mol SiCl}_4 \left(\frac{\textbf{1 mol Si}}{\textbf{1 mol SiCl}_4} \right) = \textbf{15.9 mol Si}$$

$$\textbf{15.9 mol Si} \left(\frac{\textbf{28.09 g Si}}{\textbf{1 mol Si}} \right) = \textbf{447 g Si}$$

3. If 58 moles of NH_3 are combined with 32 moles of sulfuric acid, what is the limiting reactant and how much of the excess reactant remains?

$$2\ NH_3 + H_2SO_4 → (NH_4)_2SO_4$$

$$\textbf{58 moles NH}_3 \left(\frac{\textbf{1 mole} \left(\textbf{NH}_4 \right)_2 \textbf{SO}_4}{\textbf{2 moles NH}_3} \right) = \textbf{29 moles} \left(\textbf{NH}_4 \right)_2 \textbf{SO}_4$$

$$\textbf{32 moles H}_2\textbf{SO}_4 \left(\frac{\textbf{1 mole} \left(\textbf{NH}_4 \right)_2 \textbf{SO}_4}{\textbf{1 mole H}_2\textbf{SO}_4} \right) = \textbf{32 moles} \left(\textbf{NH}_4 \right)_2 \textbf{SO}_4$$

NH_3 is the limiting reactant

$$\textbf{29 moles NH}_3 \left(\frac{\textbf{1 mole H}_2\textbf{SO}_4}{\textbf{2 moles NH}_3} \right) = \textbf{14.5 moles H}_2\textbf{SO}_4$$

32 moles H_2SO_4 - 14.5 moles H_2SO_4 = 17.5 moles H_2SO_4
remain in excess

4. What is the percent yield if 28.50 g of FeO reacts with excess CO and produces 17.841 g of Fe?

$$FeO + CO \rightarrow Fe + CO_2$$

$$28.50 \text{ g FeO} \left(\frac{1 \text{ mole FeO}}{71.850 \text{ g FeO}} \right) = 0.3967 \text{ moles FeO}$$

$$0.3967 \text{ moles FeO} \left(\frac{1 \text{ mole Fe}}{1 \text{ mole FeO}} \right) = 0.3967 \text{ moles Fe}$$

$$0.3967 \text{ moles Fe} \left(\frac{55.85 \text{ g Fe}}{1 \text{ mole Fe}} \right) = 22.15 \text{ g Fe}$$

$$\text{percent yield} = \frac{17.841 \text{ g Fe}}{22.15 \text{ g Fe}} \times 100\% = 80.55\%$$

5. What volume of 0.158 M HBr solution is required to react completely with 38.77 mL of 0.226 M Ca(OH)$_2$ in the following reaction:

$$Ca(OH)_2 + 2HBr \rightarrow CaBr_2 + 2 H_2O$$

$$38.77 \text{ mL Ca}\left(OH\right)_2 \left(\frac{0.226 \text{ mmol Ca}\left(OH\right)_2}{1 \text{ mL Ca}\left(OH\right)_2} \right) = 8.762 \text{ mmol Ca}\left(OH\right)_2$$

$$8.762 \text{ mmol Ca}\left(OH\right)_2 \left(\frac{2 \text{ mmol HBr}}{1 \text{ mmol Ca}\left(OH\right)_2} \right) = 17.52 \text{ mmol HBr}$$

$$17.52 \text{ mmol HBr} \left(\frac{1 \text{ mL HBr}}{0.158 \text{ mmol HBr}} \right) = 111 \text{ mL HBr}$$

6. How many mL of 5.44 M Sr(OH)$_2$ are required to make 100.99 mL of a 0.189 M Sr(OH)$_2$ solution? What is the molar concentration of the Sr^{2+} ions in the 0.189 M solution? What is the molar concentration of the OH$^-$ ions in the 0.189 M solution?

$$M_1 V_1 = M_2 V_2 \text{ thus } V_1 = \frac{M_2 V_2}{M_1}$$

$$V_1 = \frac{0.189 \, M \, Sr\left(OH\right)_2 \left(100.99 \text{ mL}\right)}{5.44 \, M \, Sr\left(OH\right)_2} = 3.51 \text{ mL}$$

In Sr(OH)$_2$ there is 1 Sr^{2+} ion for every 2 OH$^-$ ions.

Molar concentration of Sr^{2+} ions in 0.189 M Sr(OH)$_2$ = 0.189 M.

Molar concentration of OH$^-$ ions in 0.189 M Sr(OH)$_2$ = 0.378 M.

Module 5
Chemical Reaction Stoichiometry

Introduction

In this module we will look at several reaction stoichiometry problems. The important points to learn in this module are:
1. balancing chemical reactions
2. basic reaction stoichiometry
3. limiting reactant calculations
4. percent yield calculations
5. reactions in solution.

You will need a periodic table to calculate the molar masses in these problems.

Module 5 Key Equations & Concepts
  1. **Percent yield**

   $$\% \text{ yield} = \frac{\text{actual yield}}{\text{theoretical yield}} \times 100$$

   The percent yield formula determines the percentage of the theoretical yield formed in a reaction.
  2. *Molarity (M)*

   M = **moles solute/L solution**

   M x L = **moles** or *M* x mL = **mmol**

   The relationship of molarity and volume converts from solution concentration to moles or from a solution volume to moles of solution.

Chemical reactions symbolize how a reaction proceeds when chemical substances form new substances. Before proceeding, let's review some vocabulary:

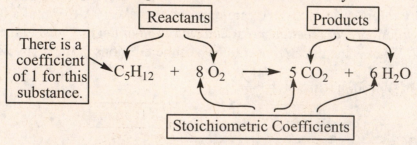

Reactants are the starting chemical species. Products result from the chemical reaction. Stoichiometric coefficients "balance" the chemical equation. Balancing ensures that equal numbers of atoms of each element are present on both sides of the reaction. Otherwise the reaction would violate the Law of Conservation of Mass.

There are numerous ways to approach balancing an equation. It should not matter where you start. However, it is often easiest to use the following steps:

TIP

1. If possible, start with an element that appears in only one compound on each side of the equation.
2. Save balancing for the last anything that is not bonded to other elements (O_2, Fe(s), etc.).
3. If the equation contains polyatomic ions, you may try balancing them as whole entities rather than each individual atom in turn.

Sample Exercises
Balancing Chemical Reactions
1. Balance this chemical reaction using the smallest whole numbers.
$$Ca(OH)_2 + H_3PO_4 \rightarrow Ca_3(PO_4)_2 + H_2O$$

Consider starting with Ca since it appears in only one compound on each side of the reaction. It is probably easiest to balance (PO_4^{3-}) as the polyatomic ion rather than individually as P and O atoms. Afterwards, only H and O are left to balance.

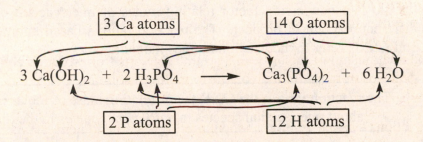

2. Balance this chemical reaction using the smallest whole numbers.
$$C_6H_{14} + O_2 \rightarrow CO_2 + H_2O$$

Start with C and balance oxygen last!

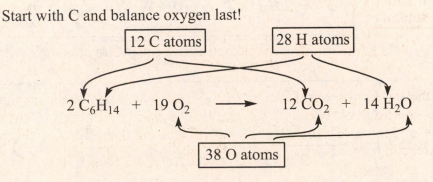

43

Simple Reaction Stoichiometry

Properly balanced chemical equations are used in chemical reaction calculations. This process is reaction stoichiometry.

3. How many moles of H$_2$ are formed from the reaction of 3.0 moles of Na with excess H$_2$O.

The correct answer is: 1.5 moles H$_2$

$$2\,Na\ +\ 2\,H_2O\ \rightarrow\ 2\,NaOH\ +\ H_2$$

INSIGHT: | The word **excess** is important. It is your clue that this problem does **not** involve a limiting reactant calculation.

The reaction ratio, derived from the balanced reaction, is 2 moles of Na consumed for every 1 mole of H$_2$ formed. Use it as a unit factor.

> Reaction Ratio

$$?\ \text{moles}\,H_2 = 3.0\ \text{moles}\,Na\left(\frac{1\,\text{mole of}\,H_2}{2\,\text{moles of}\,Na}\right) = 1.5\ \text{moles}\,H_2$$

The reaction ratio is a conversion factor relating moles of one reactant or product to moles of a different reactant or product. The reaction ratio is based upon the balanced equation. (Some texts refer to the reaction ratio as the mole ratio.)

INSIGHT: | The reaction ratio is the **ONE** factor relating mass or molar information about one reaction species to a *different species* in the reaction.

4. How many grams of H$_2$ are formed from the reaction of 11.2 grams of Na with excess H$_2$O?

The correct answer is: 0.491 g H$_2$

$$2\,Na\ +\ 2\,H_2O\ \rightarrow\ 2\,NaOH\ +\ H_2$$

Converts g of Na to moles of Na	Reaction Ratio	Converts moles of H$_2$ to g of H$_2$

$$?\ \text{grams of}\,H_2 = 11.2\ \text{g}\,Na\left(\frac{1\,\text{mole}\,Na}{22.99\,\text{g}\,Na}\right)\left(\frac{1\,\text{mole}\,H_2}{2\,\text{moles}\,Na}\right)\left(\frac{2.016\,\text{g}\,H_2}{1\,\text{mole}\,H_2}\right) = 0.491\ \text{g}\,H_2$$

Notice how the units cancel leaving grams of H$_2$.

The above problem makes the transformation from grams of one reactant, Na, to grams of one product, H₂. There is a very common set of transformations used in this calculation that appear in many reaction stoichiometry problems.

$$\text{grams of X} \rightarrow \text{moles of X} \rightarrow \text{reaction ratio} \rightarrow \text{moles of Y} \rightarrow \text{grams of Y}$$

Limiting Reactant Problems

5. *What is the maximum number of grams of H₂ formed from the reaction of 11.2 grams of Na with 9.00 grams of H₂O?*
    The correct answer is: 0.491 g of H₂

$$2\ Na + 2\ H_2O \rightarrow 2\ NaOH + H_2$$

INSIGHT: The word excess is not in this problem. Also amounts of both reactants are given. These are *clues that this is a limiting reactant problem*.

In limiting reactant problems you must perform reaction stoichiometry for each reactant amount given in the problem. This problem requires two calculations.

$$? \text{ grams of } H_2 = 11.2 \text{ g Na} \left(\frac{1 \text{ mole Na}}{22.99 \text{ g Na}} \right)\left(\frac{1 \text{ mole } H_2}{2 \text{ moles Na}} \right)\left(\frac{2.016 \text{ g } H_2}{1 \text{ mole } H_2} \right) = 0.491 \text{ g } H_2$$

$$? \text{ grams of } H_2 = 9.00 \text{ g } H_2O \left(\frac{1 \text{ mole } H_2O}{18.02 \text{ g } H_2O} \right)\left(\frac{1 \text{ mole } H_2}{2 \text{ moles } H_2O} \right)\left(\frac{2.016 \text{ g } H_2}{1 \text{ mole } H_2} \right) = 0.503 \text{ g of } H_2$$

YIELD The maximum amount formed is the *smallest* amount calculated from the reaction stoichiometry!

This calculation indicates that all 11.2 g of Na are used to produce 0.491 g of H₂. Since there is no Na left, no more H₂ can be produced, even though H₂O still remains. Once one reactant is completely consumed, no more products can be made. In this example, Na is the *limiting reactant* and H₂O is the *excess reactant*.

Percent Yield

6. *11.2 g of Na reacts with 9.00 g of H₂O and 0.400 g of H₂ is formed. What is the percent yield of the reaction?*
    The correct answer is: 81.5%

This is the actual yield.

$$2\ Na + 2\ H_2O \rightarrow 2\ NaOH + H_2$$

Key clues indicating percent yield problems are: a) amounts of both reactants given, b) an amount for a product, and c) the words percent yield.

45

$$\% \text{ yield} = \frac{\text{actual yield}}{\text{theoretical yield}} \times 100 = \frac{0.400 \text{ g}}{0.491 \text{ g}} \times 100\% = 81.5\%$$

Reaction Stoichiometry in Solution

7. How many mL of 0.250 M HCl react with 15.0 mL of 0.150 M Ba(OH)₂?
The correct answer is: 18.0 mL HCl

$$2 \text{ HCl(aq)} + Ba(OH)_2(aq) \rightarrow BaCl_2(aq) + 2 \text{ H}_2O(l)$$

INSIGHT: Key clues indicating reaction stoichiometry in solution problems are the presence of solution concentration(s) (0.250 *M* and 0.150 *M*) and volume(s) in the problem with a balanced chemical equation.

mL x *M* = millimoles Ba(OH)₂	Reaction Ratio	mmol HCl x (1/*M*) = mL HCl

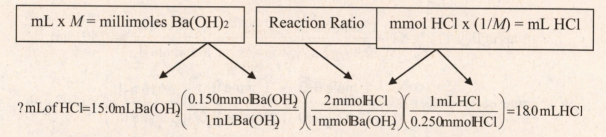

$$? \text{ mL of HCl} = 15.0 \text{ mL Ba(OH)}_2 \left(\frac{0.150 \text{ mmol Ba(OH)}_2}{1 \text{ mL Ba(OH)}_2} \right) \left(\frac{2 \text{ mmol HCl}}{1 \text{ mmol Ba(OH)}_2} \right) \left(\frac{1 \text{ mL HCl}}{0.250 \text{ mmol HCl}} \right) = 18.0 \text{ mL HCl}$$

Module 5 relates to some following Modules as shown in the graphic below.

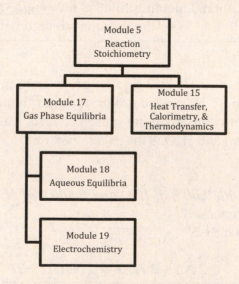

Study Tip #6

Work more conceptual problems

1. If you are memorizing the steps to solve a problem, you are in serious trouble. Understand the logical progression of step sin a problem is an improvement. Mentally "seeing" a set of steps that get you from a problem's start to finish is the best!
2. Look for commonalities in problems
   Don't be confused by problem surface features. For example, these three problems are essentially all the same problem. If you can see commonalities in problems, it will make your studying easier.

o One gallon of paint covers 650 ft^2 of wall surface. How thick is the paint once applied?

o Oil spilled on water forms a layer that is ~ 1 molecule, roughly 65 angstroms thick. If one gallon of oil is spilled on the ocean how many mile^2 of ocean surface are covered by the oil slick?

o A football field is 120.0 yards long and 53.0 yards wide. If 1.00 inch of rain falls on the field, how many pounds of rain accumulated on the field?

Module 6 Predictor Questions

The following questions may help you determine the extent you need to study this module. Questions are ranked according to ability.

    Level 1 = basic proficiency
    Level 2 = mid level proficiency
    Level 3 = high proficiency

If you can correctly answer Level 3 questions you probably do not need to spend much time on this module. If you can only answer Level 1 problems, you should review this module.

Level 1   1. Determine *all* of the reaction types that correctly classify the following reaction.

$$2\ NH_4NO_3\ (s) \rightarrow 2\ N_2\ (g) + O_2\ (g) + 4\ H_2O\ (g)$$

Level 2   2. Determine *all* of the reaction types that correctly classify the following reaction.

$$AgNO_3\ (aq) + HCl(aq) \rightarrow AgCl(s) + HNO_3(aq)$$

Level 3   3. Determine *all* of the reaction types that correctly classify the following reaction.

$$BaCO_3 \rightarrow BaO + CO_2$$

Level 2   4. Predict the products of the following reactions:
    a) $Cu(NO_3)_2 + Na_2S \rightarrow$???
    b) $CdSO_4 + H_2S \rightarrow$???
    c) $Ba(NO_3)_2 + K_2CO_3 \rightarrow$???

Level 1   5. What is the *total* ionic equation for this formula unit equation?

$$BaCl_2\ (aq) + Na_2SO_4\ (aq) \rightarrow BaSO_4(s) + 2NaCl(aq)$$

Level 2   6. What is the *net* ionic equation for the reaction of H_3PO_4 with NaOH?

Level 2   7. Write the *total* and *net* ionic equations for the reaction of solid Zn with $AgNO_3$.

Module 6 Predictor Question Solutions

1. Determine **all** of the reaction types that correctly classify the following reaction.
$$2\ NH_4NO_3\ (s) \rightarrow 2\ N_2\ (g) + O_2\ (g) + 4\ H_2O\ (g)$$

   This reaction is a decomposition reaction, an oxidation-reduction reaction, and a gas forming reaction.

2. Determine **all** of the reaction types that correctly classify the following reaction.
$$AgNO_3\ (aq) + HCl(aq) \rightarrow AgCl(s) + HNO_3(aq)$$

   This reaction is both a metathesis and a precipitation reaction.

3. Determine **all** of the reaction types that correctly classify the following reaction.
$$BaCO_3 \rightarrow BaO + CO_2$$

   This reaction is both decomposition reaction and a gas forming reaction.

4. Predict the products of the following reactions:
   - a) $Cu(NO_3)_2 + Na_2S \rightarrow$???
   - b) $CdSO_4 + H_2S \rightarrow$???
   - c) $Ba(NO_3)_2 + K_2CO_3 \rightarrow$???

   a) $Cu(NO_3)_2 + Na_2S \rightarrow NaNO_3\ (aq) + CuS(s)$
   b) $CdSO_4 + H_2S \rightarrow H_2SO_4\ (aq) + CdS(s)$
   c) $Ba(NO_3)_2 + K_2CO_3 \rightarrow KNO_3(aq) + BaCO_3(s)$

5. What is the **total** ionic equation for this formula unit equation?
$$BaCl_2\ (aq) + Na_2SO_4\ (aq) \rightarrow BaSO_4(s) + 2NaCl(aq)$$

   The *total* ionic equation is:
   $Ba^{2+}(aq) + 2Cl^-(aq) + 2Na^+(aq) + SO_4^{2-}\ (aq) \rightarrow BaSO_4\ (s) + 2Na^+(aq) + 2Cl^-(aq)$

6. What is the *net* ionic equation for the reaction of H_3PO_4 with NaOH?

   The complete molecular equation is:
   $H_3PO_4\ (aq) + 3NaOH(aq) \rightarrow Na_3PO_4(aq) + 3H_2O(l)$

   The *total* ionic equation is:
   $H_3PO_4\ (aq) + 3Na^+\ (aq) + 3OH^-(aq) \rightarrow 3Na^+\ (aq) + PO_4^{3-}\ (aq) + 3H_2O(l)$
   Note that the *weak* acid H_3PO_4 remains intact.

   The *net* ionic equation is:
   $H_3PO_4\ (aq) + 3OH^-(aq) \rightarrow PO_4^{3-}\ (aq) + 3H_2O(l)$
   (The spectator ions, Na^+, were removed.)

7. Write the *total* and *net* ionic equations for the reaction of solid Zn with $AgNO_3$.

The complete molecular equation is: $Zn(s) + 2AgNO_3 (aq) \rightarrow 2Ag(s) + Zn(NO_3)_2 (aq)$

The *total* ionic equation is:
$Zn(s) + 2Ag^+ (aq) + 2NO_3^- (aq) \rightarrow 2Ag(s) + Zn^{2+} (aq) + 2NO_3^- (aq)$

The *net* ionic equation is:
$Zn(s) + 2Ag^+ (aq) \rightarrow 2Ag(s) + Zn^{2+} (aq)$

Module 6
Chemical Reaction Types

Introduction

This module focuses on recognizing several chemical reaction types then predicting their products. The objectives of this module are:

1. how to use chemical reaction reactants to discern the reaction type
2. to predict metathesis reaction products
3. how to write total and net ionic reaction equations.

Module 6 Key Equations & Concepts

1. **Formula Unit Equations**

   Formula unit equations show all species involved in a reaction as ionic or molecular *compounds*: $KOH(aq) + HI(aq) \rightarrow KI(aq) + H_2O(l)$

2. **Total Ionic Equations**

   Total ionic equations show all ions in their ionized states. Strong acids and bases, species that ionize completely in water, are also shown as separated ions:

   $K^+(aq) + OH^-(aq) + H^+(aq) + I^-(aq) \rightarrow K^+(aq) + I^-(aq) + H_2O(l)$

   Note that all gases, solids, and liquids are left intact.

3. **Net Ionic Equations**

   To write net ionic equations, remove all spectator ions from the total ionic equation. Spectator ions are species that do not change as the reaction proceeds from reactants to products:

   $OH^-(aq) + H^+(aq) \rightarrow H_2O(l)$

Sample Exercises
Reduction-Oxidation Reactions

In reduction-oxidation reactions electrons are transferred from one species to another. Reduction cannot occur without an accompanying oxidation so these reactions are often called *redox reactions*. Your textbook has a series of rules for assigning oxidation numbers to elements in chemical species. If you do not know the rules for oxidation states, learn them now.

1. What reaction type(s) are represented by this chemical reaction?

$$2\ Na(s) + 2\ H_2O(l) \rightarrow 2\ NaOH(aq) + H_2(g)$$

The correct answer is: This is a redox reaction (later in this module we will see that it is also a combination reaction).

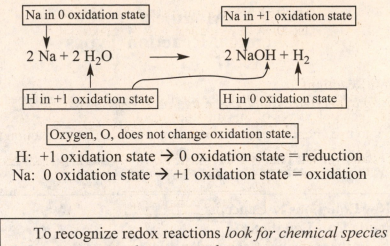

H: +1 oxidation state → 0 oxidation state = reduction

Na: 0 oxidation state → +1 oxidation state = oxidation

INSIGHT: To recognize redox reactions *look for chemical species changing oxidation states*.

Notice that Na in the above reaction is in its *elemental state* on the reactant side of the reaction and in a compound on the product side (the same is true of H). This is a big clue for a redox reaction. All species in their elemental states have oxidation states of zero, and species in compounds typically do not have oxidation states of zero. Thus, the oxidation state probably changes during the reaction!

Combination Reactions

2. What reaction type(s) are represented by this chemical reaction?

$$3\ Sr(s)\ +\ N_2(g)\ \rightarrow\ Sr_3N_2(s)$$

The correct answer is: This is both a combination and a redox reaction.

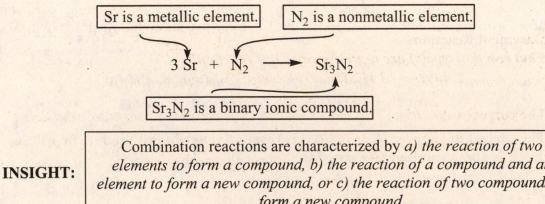

INSIGHT: Combination reactions are characterized by *a) the reaction of two elements to form a compound, b) the reaction of a compound and an element to form a new compound, or c) the reaction of two compounds to form a new compound.*

Combination reactions frequently can also be classified as another reaction type. In this example the second classification is as a redox reaction.

52

Sr: 0 → +2 oxidation state = oxidation
N: 0 → -3 oxidation state = reduction

Decomposition Reactions

3. *What reaction type(s) are represented by this chemical reaction?*
$$2\ CaO(s)\ \rightarrow\ 2\ Ca(s)\ +\ O_2(g)$$

The correct answer is: This is a decomposition reaction and a redox reaction.

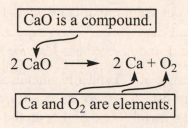

INSIGHT:	There are three types of decomposition reactions: *a) compounds decomposing into two or more elements, b) compounds decomposing into another compound and an element, and c) compounds decomposing into two simpler compounds.*

Decomposition reactions are the reverse of combination reactions. Rather than elements or compounds forming new compounds decomposition reactions break compounds into elements or less complex compounds.

As in combination reactions, decomposition reactions can also be frequently classified as other reaction types. This reaction is also a redox reaction.

Ca: +2 → 0 oxidation state = reduction
O: -2 → 0 oxidation state = oxidation

Displacement Reactions

3. *What reaction type(s) are represented by this reaction?*
$$2\ Al(s)\ +\ 3\ H_2SO_4(aq)\ \rightarrow\ Al_2(SO_4)_3(aq)\ +\ 3\ H_2(g)$$

The correct answer is: This is a displacement reaction and a redox reaction.

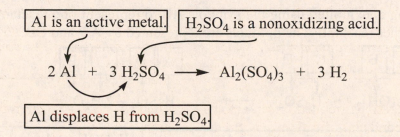

Displacement reactions involve the reaction of metals or nonmetals on the activity series with salts of less active metals or the nonoxidizing acids HCl and H_2SO_4. HNO_3 is the most common oxidizing acid.

TIP: If you are not familiar with the activity series in your text, be certain that you understand how to use it. Typically, metals higher on the list can displace metals lower on the list. *The reverse is not true.* Metals found lower on the list *cannot* displace a metal found higher on the activity series.

Metathesis Reactions

5. *What reaction type(s) are represented by this reaction?*

$$Ba(OH)_2(aq) \ + \ H_2SO_4(aq) \ \rightarrow \ BaSO_4(s) \ + \ 2\,H_2O(l)$$

The correct answer is: This is a metathesis reaction that is both an acid-base and a precipitation reaction.

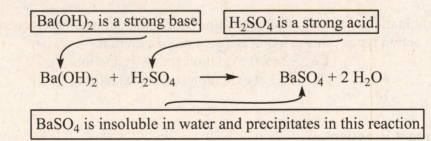

Ba(OH)₂ is a strong base. H₂SO₄ is a strong acid.

$$Ba(OH)_2 \ + \ H_2SO_4 \longrightarrow BaSO_4 + 2\,H_2O$$

BaSO₄ is insoluble in water and precipitates in this reaction.

Anion switching is exhibited using the symbols AB to represent one reactant and CD to represent the other reactant. The products are represented by AD and CB.

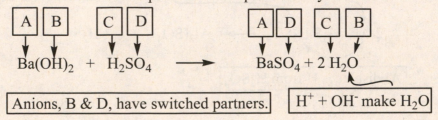

$$Ba(OH)_2 \ + \ H_2SO_4 \longrightarrow BaSO_4 + 2\,H_2O$$

Anions, B & D, have switched partners.   H⁺ + OH⁻ make H₂O

In simple acid-base reactions as an acid reacts with a base, a salt ($BaSO_4$ in this case) and water (if the base is a hydroxide) are formed. Water is formed by the reaction of H^+ with OH^- from the anion partner switch.

INSIGHT: Precipitation reactions are characterized by *the formation of an insoluble compound in water*.

△ TIP — You must understand and use the solubility rules from your textbook to recognize a precipitation reaction since the phases of the product compounds are not frequently given, as illustrated in the exercises below.

Predicting Products of Metathesis Reactions
6. *What are the products of this chemical reaction?*

$$Sr(OH)_2(aq) \ + \ Fe(NO_3)_3(aq) \rightarrow ??? + ???$$

The correct answer is: $Sr(NO_3)_2$ and $Fe(OH)_3$

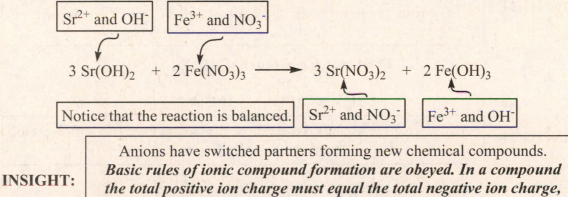

Sr^{2+} and OH^- | Fe^{3+} and NO_3^-

$$3 \ Sr(OH)_2 \ + \ 2 \ Fe(NO_3)_3 \longrightarrow 3 \ Sr(NO_3)_2 \ + \ 2 \ Fe(OH)_3$$

Notice that the reaction is balanced. | Sr^{2+} and NO_3^- | Fe^{3+} and OH^-

INSIGHT: Anions have switched partners forming new chemical compounds. *Basic rules of ionic compound formation are obeyed. In a compound the total positive ion charge must equal the total negative ion charge, resulting in neutral compound formation.*

Total and Net Ionic Equations
Net ionic equations are very helpful to focus us on the essential reaction parts. For example, net ionic equations make assigning oxidation numbers much easier.

7. *Write the total ionic and net ionic equations for this reaction.*

$$Ba(OH)_2(aq) \ + \ 2 \ HCl(aq) \ \rightarrow \ BaCl_2(aq) \ + \ 2 \ H_2O(l)$$

The correct total ionic equation is:

$$Ba^{2+}(aq) + 2 \ OH^-(aq) + 2 \ H^+(aq) + 2 \ Cl^-(aq) \rightarrow Ba^{2+}(aq) + 2 \ Cl^-(aq) + 2 \ H_2O(l)$$

The correct net ionic equation is:

$$2 \ OH^-(aq) + 2 \ H^+(aq) \rightarrow 2 \ H_2O(l)$$
$$\text{or}$$
$$OH^-(aq) + H^+(aq) \rightarrow H_2O(l)$$

These are very difficult problems if you are not readily familiar with the solubility rules.

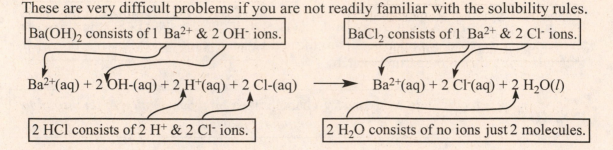

$Ba(OH)_2$ consists of 1 Ba^{2+} & 2 OH^- ions.

$BaCl_2$ consists of 1 Ba^{2+} & 2 Cl^- ions.

$Ba^{2+}(aq) + 2\ OH^-(aq) + 2\ H^+(aq) + 2\ Cl^-(aq) \longrightarrow Ba^{2+}(aq) + 2\ Cl^-(aq) + 2\ H_2O(l)$

2 HCl consists of 2 H^+ & 2 Cl^- ions.

2 H_2O consists of no ions just 2 molecules.

The 2 before the OH^- comes from the subscript 2 in $Ba(OH)_2$. The 2's before H^+ and Cl^- come from the stoichiometric coefficient 2.

△ TIP   If there are both a subscript and a coefficient, multiply them together to determine the number of ions present.

INSIGHT: *Spectator ions do not change from reactant to product.* Ba^{2+} and Cl^- are spectator ions in this reaction. Once the correct total ionic equation is written, *removal of the spectator ions, Ba^{2+} and Cl^-, leaves the correct net ionic equation.*

$$2\ OH^-(aq) + 2\ H^+(aq) \rightarrow 2\ H_2O(l)$$
or
$$OH^-(aq) + H^+(aq) \rightarrow H_2O(l)$$

INSIGHT: As discussed in Module 5, we reduce the stoichiometric coefficients to their smallest whole numbers.

8. *Write the total and net ionic equations for this reaction.*
    $$NaOH(aq) + CH_3COOH(aq) \rightarrow NaCH_3COO(aq) + H_2O(l)$$

The correct total ionic equation is:
$$Na^+(aq) + OH^-(aq) + CH_3COOH(aq) \rightarrow Na^+(aq) + CH_3COO^-(aq) + H_2O(l)$$

The correct net ionic equation is:
$$OH^-(aq) + CH_3COOH(aq) \rightarrow CH_3COO^-(aq) + H_2O(l)$$

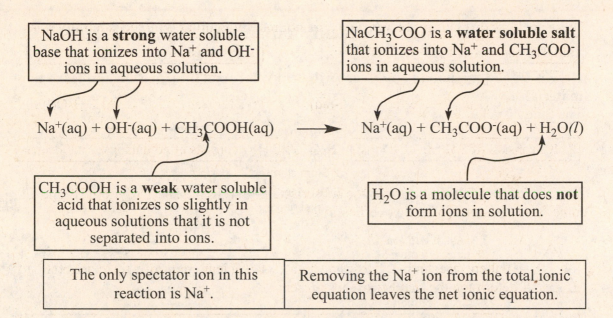

NaOH is a **strong** water soluble base that ionizes into Na^+ and OH^- ions in aqueous solution.

NaCH$_3$COO is a **water soluble salt** that ionizes into Na^+ and CH_3COO^- ions in aqueous solution.

$$Na^+(aq) + OH^-(aq) + CH_3COOH(aq) \longrightarrow Na^+(aq) + CH_3COO^-(aq) + H_2O(l)$$

CH$_3$COOH is a **weak** water soluble acid that ionizes so slightly in aqueous solutions that it is not separated into ions.

H$_2$O is a molecule that does **not** form ions in solution.

The only spectator ion in this reaction is Na^+.

Removing the Na^+ ion from the total ionic equation leaves the net ionic equation.

$$OH^- (aq) + CH_3COOH(aq) \rightarrow CH_3COO^-(aq) + H_2O(l)$$

Notice that the net ionic equation tells us that a strong base, hydroxide ion, reacts with the weak acid, acetic acid, to form the acetate ion and water.

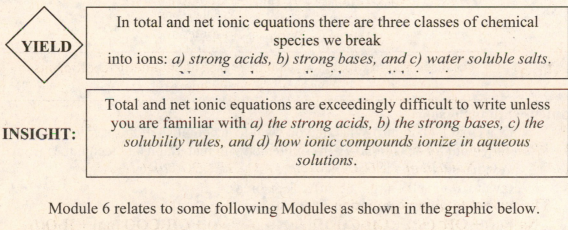

YIELD

In total and net ionic equations there are three classes of chemical species we break into ions: *a) strong acids, b) strong bases, and c) water soluble salts.*

INSIGHT:

Total and net ionic equations are exceedingly difficult to write unless you are familiar with *a) the strong acids, b) the strong bases, c) the solubility rules, and d) how ionic compounds ionize in aqueous solutions.*

Module 6 relates to some following Modules as shown in the graphic below.

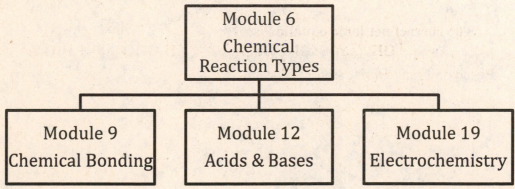

Module 6
Chemical
Reaction Types

Module 9
Chemical Bonding

Module 12
Acids & Bases

Module 19
Electrochemistry

Study Tip #7

Here is a good method to study that will help you retain information and learn new material.

The Study Cycle

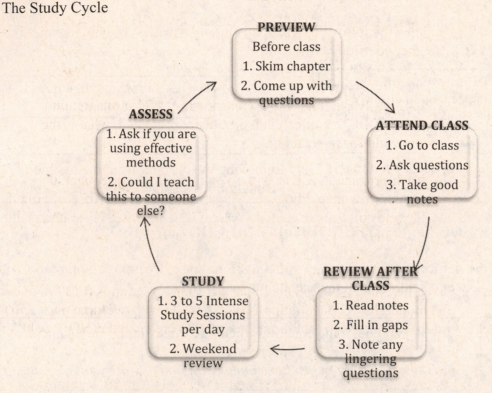

Intense Study Sessions
1. Set a Goal – (1-2 min) What do you want to accomplish in this session?
2. Study with a Focus – (30-50 min) Interact with material using techniques from previous study tips.
3. Reward Yourself – (10-15 min) Take a break.
4. Review (5 min) – Reflect on what you have done.

Personal Communication – S. Y. Mcguire
Based on Frank Christ's PLRS system.

Practice Test Two
Modules 4-6

Level 1   1. What is the mass of the oxygen atoms in 42.7 g of CH_3COOH?

Level 1   2. Balance the following reaction using the smallest possible whole
            number coefficients.
            ___P_4O_{10} + ___H_2O → ___H_3PO_4

Level 1   3. A 28.42 g silver nitrate sample reacts with 14.00 g of calcium
            chloride. If 10.72 g of calcium nitrate are produced, what is the
            percent yield for the reaction?

Level 3   4. Hydrochloric acid, HCl, is formed by the following sequential
            reactions. How many moles of HCl are formed from 105 g of H_2O if
            the % yields of steps one and two are 67.2% and 86.9%, respectively?
            $2 H_2O \rightarrow 2 H_2 + O_2$ (67.2% yield)
            $H_2 + Cl_2 \rightarrow 2 HCl$ (86.9% yield)

Level 2   5. In an $Fe(NO_3)_3$ solution the molar Fe^{3+}concentration is 0.150 M and
            the molar NO_3^- concentration is 0.450 M. To what volume must 250.0
            mL of this $Fe(NO_3)_3$ solution be diluted to create an $Fe(NO_3)_3$ solution
            having a 0.0850 M concentration?

Level 1   6. Excess $AlCl_3$ reacts with 52.3 mL of 0.500 M $AgNO_3$ according to the
            following reaction.
            $AlCl_3 + 3 AgNO_3 \rightarrow 3 AgCl + Al(NO_3)_3$
            What is the final solution volume if the resulting $Al(NO_3)_3$
            concentration is 0.0673 M?

Level 1   7. Classify using *all applicable* reaction types the reactions in questions
            4 and 6.

Level 1   8. Write both the *total* and the *net* ionic equations for the reaction of
            chloric acid with strontium hydroxide.

Module 7 Predictor Questions

The following questions may help you determine the extent you need to study this module. Questions are ranked according to ability.

      Level 1 = basic proficiency

      Level 2 = mid level proficiency

      Level 3 = high proficiency

If you can correctly answer Level 3 questions you probably do not need to spend much time on this module. If you can only answer Level 1 problems, you should review this module.

Level 1   1. What is the ground state electron configuration of tellurium, $_{52}$Te?

Level 2   2. What is the principal quantum number for the ***valence*** electrons of each element?
          a) K
          b) P
          c) Mn

Level 2   3. Choose the set of quantum numbers that are **NOT** correct for any electron in the ground state configuration of Si.
          a) $n = 3, \ell = 2, m_\ell = -1, m_s = +1/2$
          b) $n = 2, \ell = 1, m_\ell = -1, m_s = -1/2$
          c) $n = 3, \ell = 0, m_\ell = 0, m_s = +1/2$
          d) $n = 2, \ell = 0, m_\ell = 0, m_s = -1/2$
          e) $n = 1, \ell = 0, m_\ell = 0, m_s = -1/2$

Level 3   4. Answer the following questions regarding Fe.
          a) How many d electrons are present in the ground state electron configuration of Fe?
          b) What is the value of the ***n*** quantum number for the d electrons in Fe?
          c) How many of the d electrons in Fe are *paired*?
          d) How many of the d electrons in Fe are *unpaired*?

Level 3   5. What is the maximum number of electrons in a Au atom that are described by the following quantum number sets?
          a) $n = 3$
          b) $n = 3, \ell = 1$
          c) $n = 5, \ell = 2, m_\ell = -1$
          d) $n = 4, \ell = 3$
          e) $n = 2, \ell = 0, m_\ell = 0, m_s = -1/2$

Level 3   6. The spatial orientation of an atomic orbital is designated by which of
                the four quantum numbers?

Level 3   7. What is the value of the angular momentum quantum number for each
                of these atomic orbitals?
                a) s
                b) d
                c) p
                d) f

Module 7 Predictor Question Solutions

1. What is the ground state electron configuration of tellurium, $_{52}$Te?

    Te: $1s^2 2s^2 2p^6 3s^2 3p^6 4s^2 3d^{10} 4p^6 5s^2 4d^{10} 5p^4$ or [Kr] $5s^2 4d^{10} 5p^4$

2. What is the principal quantum number for the *valence* electrons of each element?
    a) K   **$n = 4$**
    b) P   **$n = 3$**
    c) Mn **$n = 3$**

3. Choose the set of quantum numbers that are **NOT** correct for any electron in the ground state configuration of Si.
    a) $n = 3, \ell = 2, m_\ell = -1, m_s = +1/2$
    b) $n = 2, \ell = 1, m_\ell = -1, m_s = -1/2$
    c) $n = 3, \ell = 0, m_\ell = 0, m_s = +1/2$
    d) $n = 2, \ell = 0, m_\ell = 0, m_s = -1/2$
    e) $n = 1, \ell = 0, m_\ell = 0, m_s = -1/2$

    Statement a) is untrue. The combination of quantum numbers $n = 3$ and $\ell = 2$ indicates an electron located in the d block of the fourth row of the periodic table. Since Si is located in the p block of the third row, this electron configuration is not found in the Si ground state configuration.

4. Answer the following questions regarding Fe.
    a) How many d electrons are present in the ground state electron configuration of Fe?
    b) What is the value of the *n* quantum number for the d electrons in Fe?
    c) How many of the d electrons in Fe are *paired*?
    d) How many of the d electrons in Fe are *unpaired*?

    The ground state electron configuration for Fe is: [Ar]$4s^2 3d^6$
    a) There are six d electrons.
    b) $n = 3$
    c) and d) If Hund's rule is obeyed, then the d electrons must be placed into the d orbitals such that each orbital contains one electron before any orbital contains two electrons. Thus, there are two paired electrons and four unpaired electrons.

5. What is the maximum number of electrons in a Au atom that are described by the following quantum number sets?
    a) $n = 3$ **18 electrons**
    b) $n = 3, \ell = 1$ **6 electrons**
    c) $n = 5, \ell = 2, m_\ell = -1$ **2 electrons**
    d) $n = 4, \ell = 3$ **14 electrons**
    e) $n = 2, \ell = 0, m_\ell = 0, m_s = -1/2$ **1 electron**

6. The spatial orientation of an atomic orbital is designated by which of the four quantum numbers?

The orbital spatial orientation is given by m_ℓ.

7. What is the value of the angular momentum quantum number for each of the following atomic orbitals?

a) s	$\boldsymbol{\ell = 0}$	
b) d	$\boldsymbol{\ell = 2}$	
c) p	$\boldsymbol{\ell = 1}$	
d) f	$\boldsymbol{\ell = 3}$	

Module 7
Electronic Structure of Atoms

Introduction

This module describes the meaning of quantum numbers and how to ssign them to electrons in an atom. The goals of this module are:

1. how to determine the quantum numbers for the electrons in an atom
2. how to discern the correct ground state electronic structure of an element from the periodic table
3. how to write the entire set of quantum numbers for an atom.

You will need access to a periodic table to work on the sample exercises in this module.

Module 7 Key Equations & Concepts

1. **Principal quantum number**

   Represented by the symbol **n**, this quantum number describes the main energy level of an atom.

   $$n = 1, 2, 3, 4, 5, 6, \ldots \infty$$

2. **Angular momentum quantum number**

   Represented by the symbol ℓ the angular momentum quantum number describes atomic orbital shapes as well as the region of space occupied by those electrons. Allowed values of ℓ are dependent on the n value. Each value of ℓ corresponds to a specific type of orbital.

   $$\ell = 0, 1, 2, 3, 4, \ldots (n-1)$$
   $$\ell = s, p, d, f, g, \ldots (n-1)$$

3. **Magnetic quantum number**

   Represented by the symbol m_ℓ, the magnetic quantum number describes the number of possible atomic orbitals for each ℓ value.

   $$m_\ell = -\ell, -\ell+1, -\ell + 2, \ldots, 0, \ldots \ell-2, \ell-1, \ell$$

4. **Spin quantum number**

   Represented by the symbol m_s, the spin quantum number describes the relative magnetic orientation of the electrons in an atom. It also defines the maximum number of electrons that can occupy one orbital.

   $$m_s = +1/2 \text{ or } -1/2$$

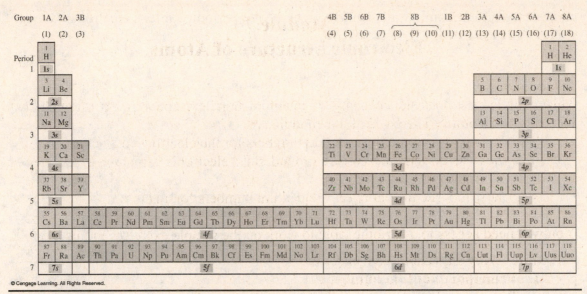

Sample Exercises

Principal Quantum Number

1. What is the value of the principal quantum number, n, for the valence electrons in a Sr atom?

The correct answer is: $n = 5$

> **INSIGHT:** Sr is on the 5th row of the periodic table. All of the elements on the 5th row, except the transition metals, have n = 5. The value of n equals the period number except for transition metals, lanthanides, and actinides.

2. What is the value of the n quantum number for the valence electrons in a Zr atom?

The correct answer is: $n = 4$

> **INSIGHT:** Zr is a transition metal on the 5th row of the periodic table. ***d electrons in d-Transition metals have an n value one less than their periodic table period number.*** Zr, a 5th row transition metal has n = 5-1 = 4 for its d electrons.

Orbital Angular Momentum Quantum Number

3. What is the value of the orbital angular momentum quantum number, ℓ, for the valence electrons in a Sr atom?

The correct answer is: ℓ = 0

Valence electrons in Sr have n = 5, so ℓ has values of **0, 1, 2, 3, 4**. Remember these ℓ values correspond to **s, p, d, f, and g** orbitals, respectively. The two electrons that make Sr different from Ar are in the s block of the periodic table. s electrons have an ℓ value of 0.

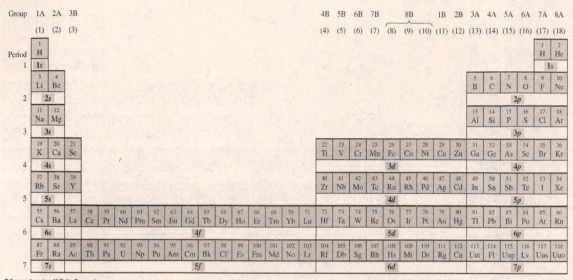

© Cengage Learning. All Rights Reserved.

TIPS

Successive elements on the periodic table have one additional proton and one electron. This fact along with the knowledge that every additional orbital holds at most two electrons is a mnemonic to decipher which part of the periodic table represents each orbital type.

- The two columns of the periodic table for the valence shell electrons in alkali and alkaline earth metals represent <u>one s orbital</u> per period.
- The six electrons that appear in boron through neon in row two are in <u>three p orbitals</u>. That is also true for periods 3 through 7.
- The ten d-electrons possible for the transition metal elements fill <u>five d-orbitals</u>.
- The fourteen elements in the lanthanide and actinide periods fill <u>seven f orbitals</u>.

4. *What is the value of the orbital angular momentum quantum number, ℓ, for the electrons in a Zr atom having different ℓ values from a Sr atom?*
   The correct answer is: $\ell = 2 = d$ electrons

For Zr which has n = 5, ℓ may be **0, 1, 2, 3, or 4** (corresponding to **s, p, d, f, or g** orbitals).

66

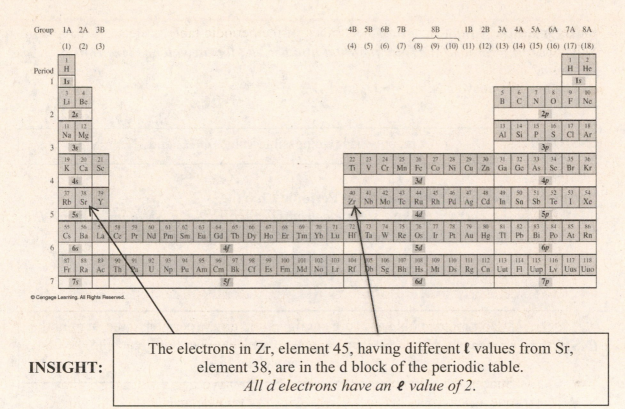

Group 1A 2A 3B   4B 5B 6B 7B ⌐8B⌐ 1B 2B 3A 4A 5A 6A 7A 8A
(1) (2) (3)   (4) (5) (6) (7) (8) (9) (10) (11) (12) (13) (14) (15) (16) (17) (18)

© Cengage Learning. All Rights Reserved.

INSIGHT: The electrons in Zr, element 45, having different ℓ values from Sr, element 38, are in the d block of the periodic table.
All d electrons have an ℓ value of 2.

Magnetic Quantum Number

5. What is the value of the magnetic quantum number, m_ℓ, for the valence electrons in a Sr atom?

The correct answer is: $m_\ell = 0$

INSIGHT: Sr valence electrons, are s electrons having ℓ = 0. The magnetic quantum number has any integer value from -ℓ to +ℓ. If ℓ = 0, then the only possible value of m_ℓ is 0.

6. What is the value of the magnetic quantum number, m_ℓ, for the electrons in a Zr atom having different ℓ values from a Sr atom?

The correct answer is: $m_\ell = -2, -1, 0, +1, +2$

Electrons in a Zr atom having different ℓ values from Sr are d electrons which always have an ℓ = 2. For ℓ = 2 there are five values possible values of m_ℓ (**-2, -1, 0, +1, +2** is five different values; see the concepts box at the beginning of the module if you do not understand how these values were derived), indicating there are five different d orbitals.

INSIGHT: You must think of quantum numbers as labels rather than numbers.

Spin Quantum Number

7. *What is the value of the spin quantum number, m_s, for the valence electrons in a Sr atom?*

   The correct answer is: $m_s = +1/2$ and $-1/2$

INSIGHT:	m_s only has two possible values, $+1/2$ and $-1/2$.

Determine Electronic Structure from Periodic Chart

8. *What is the correct ground state electronic structure for a Sr atom? Write the structure in both orbital notation and simplified (or spdf) notation.*

   The correct answer is: [Kr] $\uparrow\downarrow$ or [Kr] $5s^2$
   $5s$

INSIGHT:	To find the noble gas core configuration on the periodic table, start at the element of interest then decrease atomic number until reaching the first noble gas.

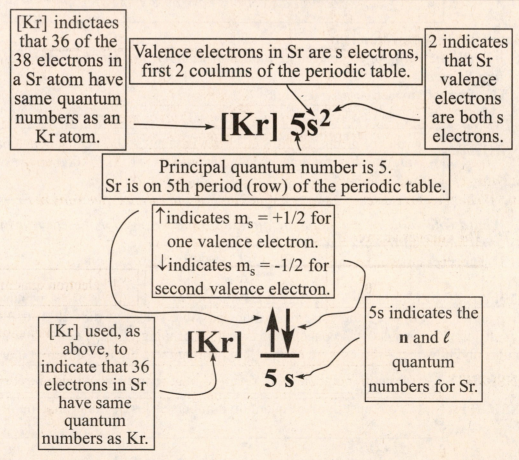

[Kr] indictaes that 36 of the 38 electrons in a Sr atom have same quantum numbers as an Kr atom.

Valence electrons in Sr are s electrons, first 2 coulmns of the periodic table.

2 indicates that Sr valence electrons are both s electrons.

$[Kr]\ 5s^2$

Principal quantum number is 5.
Sr is on 5th period (row) of the periodic table.

$\uparrow$ indicates $m_s = +1/2$ for one valence electron.
$\downarrow$ indicates $m_s = -1/2$ for second valence electron.

[Kr] used, as above, to indicate that 36 electrons in Sr have same quantum numbers as Kr.

[Kr] $\uparrow\downarrow$
$5\ s$

5s indicates the **n** and ℓ quantum numbers for Sr.

9. *What is the correct ground state electronic structure of the Zr atom? Write the structure in both orbital notation and simplified (or spdf) notation.*

The correct answer is: [Kr] $\underset{5s}{\uparrow\downarrow}$ $\underset{4d}{\uparrow\ \ \uparrow}$ __ __ __ or [Kr] $5s^2\,4d^2$

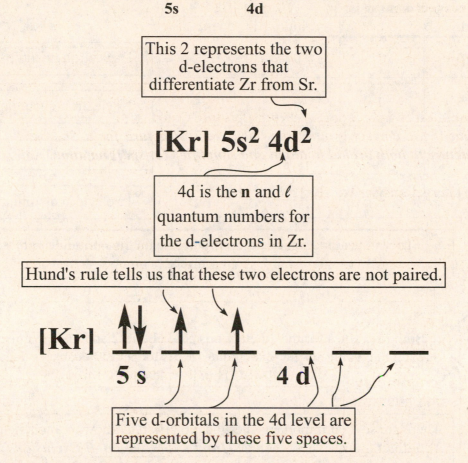

This 2 represents the two d-electrons that differentiate Zr from Sr.

$$[Kr]\ 5s^2\ 4d^2$$

4d is the **n** and ℓ quantum numbers for the d-electrons in Zr.

Hund's rule tells us that these two electrons are not paired.

[Kr] $\underset{5\,s}{\uparrow\downarrow}$ $\uparrow$ $\uparrow$ __ __ __ $\;\;4\,d$

Five d-orbitals in the 4d level are represented by these five spaces.

Writing Quantum Numbers

10. *Write the correct set of quantum numbers for the valence electrons in Sr.*

The correct answer is:

n	ℓ	m_ℓ	m_s	
5	0	0	+1/2	1st electron quantum numbers
5	0	0	-1/2	2nd electron quantum numbers

INSIGHT: If both m_s numbers were reversed, that answer would also be correct.

11. *Write the correct set of quantum numbers for the valence electrons in Zr.*

The correct answer is:

n	ℓ	m_ℓ	m_s	
5	0	0	+1/2	1st electron quantum numbers
5	0	0	-1/2	2nd electron quantum numbers
4	2	-2	+1/2	3rd electron quantum numbers
4	2	-1	+1/2	4th electron quantum numbers

INSIGHT: Strictly speaking, m_ℓ could be any two of the five possible values and be correct.

 **CAUTION** Both m_s values for the 4d electrons could be -1/2 and also be correct. But having one value +1/2 and the other -1/2 is incorrect because it does not obey Hund's rule.

YIELD There are several important rules that you need to know to understand electron configurations. These include *Hund's rule,* the *Pauli Exclusion Principle,* and the *Aufbau Principle*. Be certain that you understand these rules. The most important thing to learn from this module is how to get the correct electron configuration for an element using the periodic table.

Module 7 relates to some following Modules as shown in the graphic below.

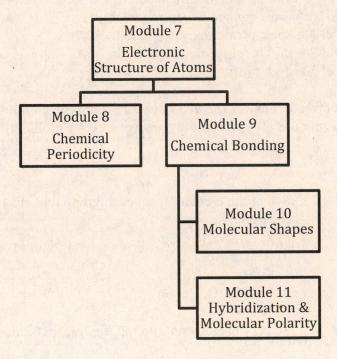

Module 8 Predictor Questions

The following questions may help you determine the extent you need to study this module. Questions are ranked according to ability.

       Level 1 = basic proficiency

       Level 2 = mid level proficiency

       Level 3 = high proficiency

If you can correctly answer Level 3 questions you probably do not need to spend much time on this module. If you can only answer Level 1 problems, you should review this module.

Level 1

1. Select the element from each group that has the **largest** electronegativity.
    a) S, Zn, Na, Te, Cu
    b) Al, Cr, Rb, Li, N
    c) K, Sb, Au, Cl, Ba
    d) Pd, Mg, O, Po, Sr

Level 1

2. Select the element with the **highest** first ionization energy.
    B, Al, Ga, In, Tl

Level 1

3. Which element has the **smallest** atomic radius?
    Mo, Au, Bi, In, Te

Level 3

4. Rank the following elements in order of **increasing** first ionization energy: C, B, N, O

Level 3

5. Which of these elements has the **most negative** electron affinity?
    As, Al, K, Se, Sn

Level 2

6. Arrange these ions in order of **increasing** ionic radii.
    a) Al^{3+}, Na^+, Mg^{2+}
    b) F^-, N^{3-}, O^{2-}
    c) F^-, Na^+, Mg^{2+}, O^{2-}

Module 8 Predictor Question Solutions

1. Select the element from each group that has the *largest* electronegativity
   a) S, Zn, Na, Te, Cu
   b) Al, Cr, Rb, Li, N
   c) K, Sb, Au, Cl, Ba
   d) Pd, Mg, O, Po, Sr

   Electronegativity increases going up a periodic table column and from left to right across a period.
   a) S
   b) N
   c) Cl
   d) O

2. Select the element with the *highest* first ionization energy.
   B, Al, Ga, In, Tl

   First ionization energies increase going up a periodic table column and from left to right across a period. B has the greatest first ionization energy of these elements.

3. Which element has the *smallest* atomic radius?
   Mo, Au, Bi, In, Te

   Atomic radius increases going down a periodic table column and from right to left across a period, so the atom in this group with the smallest atomic radius is Te.

4. Rank the following elements in order of *increasing* first ionization energy.
   C, B, N, O

   The same trend described in number 2 is followed. However, N has a greater first ionization energy than O because its valence electron configuration contains 3 p electrons (half-filled p subshell). Therefore, N has a more energetically favorable configuration, and it is more difficult to remove an electron.
   $$B < C < O < N$$

5. Which of the following elements has the *most negative* electron affinity?
   As, Al, K, Se, Sn

   Electron affinity becomes more negative going up a periodic table column and from left to right across a period. Thus, Se has the most negative electron affinity of this group.

73

6. Arrange these ions in order of *increasing* ionic radii.
   a) Al^{3+}, Na^+, Mg^{2+}
   b) F^-, N^{3-}, O^{2-}
   c) F^-, Na^+, Mg^{2+}, O^{2-}

For an isoelectronic series, ionic radius decreases with increasing atomic number. Keep in mind that cations have smaller radii than their neutral parent atoms while anions have larger radii than their neutral parent atoms.

   a) $Al^{3+} < Mg^{2+} < Na^+$
   b) $F^- < O^{2-} < N^{3-}$
   c) $Mg^{2+} < Na^+ < F^- < O^{2-}$

Module 8
Chemical Periodicity

Introduction

Several important elemental properties are based upon the electronic structures of the elements. Based on some simple rules, we can predict variations of these properties solely upon elemental position on the periodic chart. The primary goals of this module are to understand the periodic properties associated with:

1. electronegativity
2. ionization energy
3. electron affinity
4. atomic radii
5. ionic radii.

It will help to have a periodic table with you as you work on this module. You must learn to associate these properties with periodic table trends.

Module 8 Key Equations & Concepts

1. Electronegativity

Electronegativity is a relative measure of an element's ability to attract electrons to itself in a chemical bond. This property helps us determine the likelihood of ionic or covalent bond formation as well as molecular polarity.

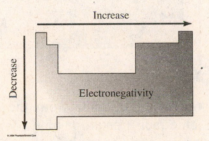

2. Ionization energy

Ionization energy is the energy required to remove an electron from an atom or ion. This property is an important indicator of an element's likelihood of forming positive ions. Elements with several electrons have a 1st ionization energy, 2nd ionization energy, and so forth until all of that element's electrons have been removed. Note the similarity to the trend in electronegativity.

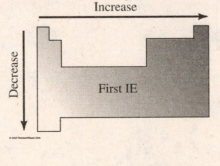

---

3. Electron affinity

Electron affinity is the energy absorbed when an electron is added to an isolated gaseous atom. Electron affinity helps us understand which elements likely form negative ions. Electron affinity has the most irregular periodic trend of the properties discussed in this module.

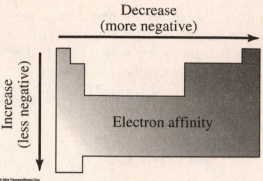

4. Atomic radii

Atomic radii are the measured distances from the center of the atom to its outer electrons. Atomic radii help us predict solid state structures of elements.

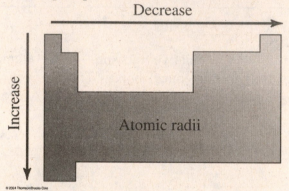

5. Ionic radii

This property is the measured distance from the center of an ion to its outer electrons. There are ionic radii trends for both positive and negative ions. These trends will help us determine ionic bond strengths. *Cations are always smaller than their parent atoms and anions are always larger than their parent atoms.*

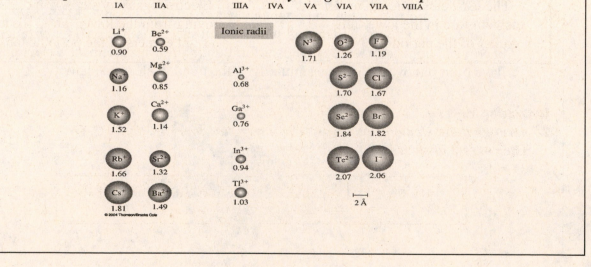

Electronegativity
1. Arrange these elements in order of increasing electronegativity: O, Ca, Si, Cs
The correct answer is: Cs < Ca < Si < O

The most electronegative elements are in the upper right corner of the periodic table.

Electronegativities of the Elements

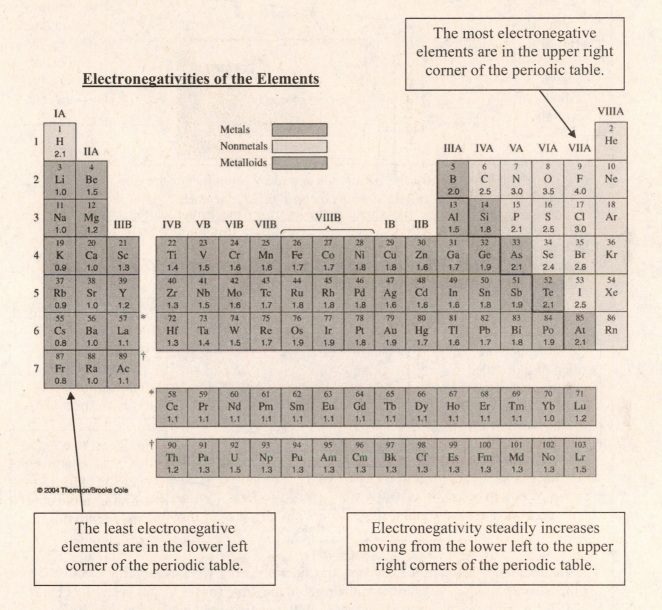

© 2004 Thomson/Brooks Cole

The least electronegative elements are in the lower left corner of the periodic table.

Electronegativity steadily increases moving from the lower left to the upper right corners of the periodic table.

Ionization Energy
2. Arrange these elements by increasing ionization energies: F, N, C, O
The correct answer is: C < O < N < F

First ionization energies increase steadily from the alkali metals to the noble gases.

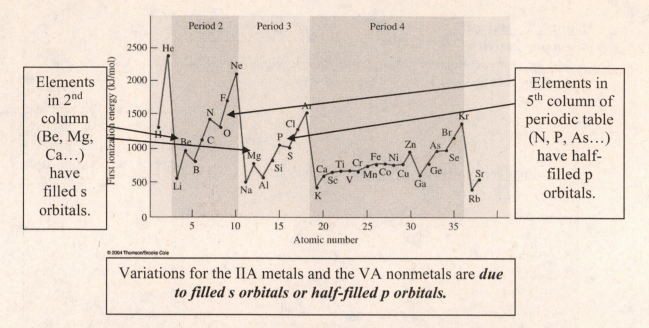

Elements in 2nd column (Be, Mg, Ca…) have filled s orbitals.

Elements in 5th column of periodic table (N, P, As…) have half-filled p orbitals.

Variations for the IIA metals and the VA nonmetals are *due to filled s orbitals or half-filled p orbitals.*

Electron Affinity

3. Arrange these elements by increasing electron affinity: F, N, C, O

The correct answer is: **F < O < C < N**

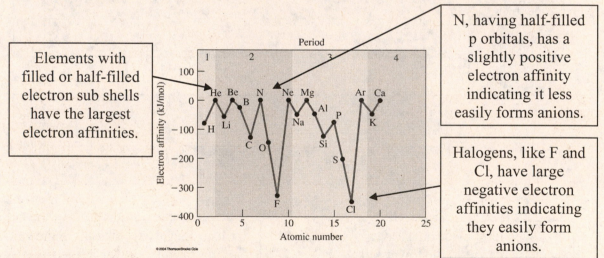

Elements with filled or half-filled electron sub shells have the largest electron affinities.

N, having half-filled p orbitals, has a slightly positive electron affinity indicating it less easily forms anions.

Halogens, like F and Cl, have large negative electron affinities indicating they easily form anions.

The generic trend shown in the concepts box does not work when comparing electron affinities for C and N. N is farther to the right in the same row as C, but because N has a half-filled p orbital its electron affinity is slightly positive and greater than that of C.

Atomic Radii

4. Arrange these elements by increasing atomic radii: F, Ga, S, Rb

The correct answer is: **F < S < Ga < Rb**

78

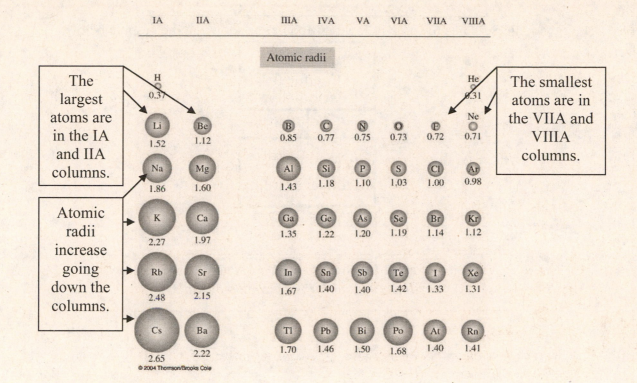

The largest atoms are in the IA and IIA columns.

Atomic radii increase going down the columns.

The smallest atoms are in the VIIA and VIIIA columns.

© 2004 Thomson/Brooks Cole

Ionic Radii

5. *Arrange these ions by increasing ionic radii:* S^{2-}, Cl^-, Mg^{2+}, Al^{3+}
   The correct answer is $Al^{3+} < Mg^{2+} < Cl^- < S^{2-}$

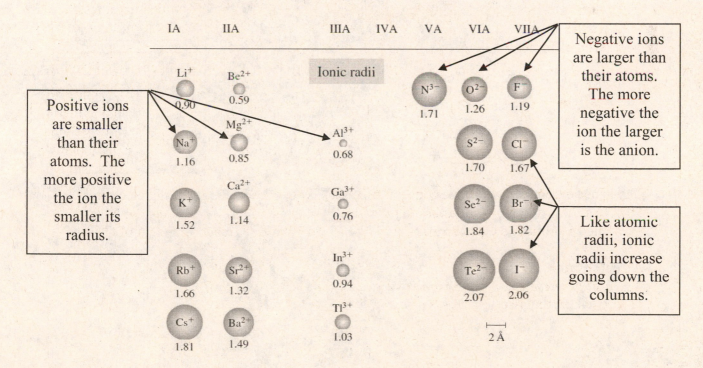

Positive ions are smaller than their atoms. The more positive the ion the smaller its radius.

Negative ions are larger than their atoms. The more negative the ion the larger is the anion.

Like atomic radii, ionic radii increase going down the columns.

Module 8 relates to some following Modules as shown in the graphic below.

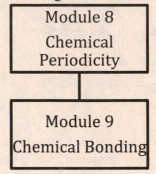

80

Module 9 Predictor Questions

The following questions may help you determine the extent you need to study this module. Questions are ranked according to ability.

  Level 1 = basic proficiency
  Level 2 = mid level proficiency
  Level 3 = high proficiency

If you can correctly answer Level 3 questions you probably do not need to spend much time on this module. If you can only answer Level 1 problems, you should review this module.

Level 1   1. Chlorine is ***most likely*** to form an ionic compound with which of these elements?   F, O, C, N, Li

Level 3   2. Choose all of the ***ionic*** compounds from the list below.
            K_3N, $CaBr_2$, Li_2O, HI, CF_4, OBr_2

Level 3   3. Choose all of the ***covalent*** compounds from the list below.
            $Ca(OH)_2$, Li_3N, Sr_3N_2, CO_2, NI_3, CBr_4

Level 2   4. Name the following ionic compounds then determine how many ions are present in one formula unit of each.
            a) $AlPO_4$
            b) $Mg(NO_3)_2$
            c) Na_2CO_3

Level 1   5. Draw the Lewis dot structures of the following atoms: B, P, K, and S

Level 1   6. Which chemical formula is ***incorrect***?
            $SrBr_2$, K_2S, $MgSe$, $CsCl_2$, Al_2O_3

Level 1   7. Which compound involves ***both*** ionic and covalent bonding?
            Cl_2, Na_2SO_4, KCl, HF, HCN

Level 1   8. Which molecule ***does not*** have a dipole moment (which is nonpolar?)
            $BrCl$, ClF, BrF, O_2, ICl

Module 9 Predictor Question Solutions

1. Chlorine is *most likely* to form an ionic compound with which of these elements?
   F, O, C, N, Li

   Chlorine is a nonmetal that can form the Cl⁻ anion. It is most likely to form an ionic compound with a metal such as Li which forms the Li⁺ cation.

2. Choose all of the *ionic* compounds from the list below.
   K_3N, $CaBr_2$, Li_2O, HI, CF_4, OBr_2

   Ionic compounds are formed between a metal and a nonmetal OR a metal and a polyatomic anion. The ionic compounds in this list are: K_3N, $CaBr_2$, and Li_2O. Note that HI is NOT ionic since H and I are both nonmetals.

3. Choose all of the *covalent* compounds from the list below.
   $Ca(OH)_2$, Li_3N, Sr_3N_2, CO_2, NI_3, CBr_4

   Covalent compounds form between two or more nonmetals. The covalent compounds in this list are: CO_2, NI_3, and CBr_4.

4. Name the following ionic compounds then determine how many ions are present in one formula unit of each.
   a) $AlPO_4$
   b) $Mg(NO_3)_2$
   c) Na_2CO_3

   a) $AlPO_4$ aluminum phosphate; two ions
   b) $Mg(NO_3)_2$; magnesium nitrate; three ions
   c) Na_2CO_3; sodium carbonate; three ions

5. Draw the Lewis dot structures of the following atoms: B, P, K, and S

   B⋅   ⋅P̈⋅   K̇   ⋅S̈:

6. Which chemical formula is *incorrect*?
   $SrBr_2$, K_2S, MgSe, $CsCl_2$, Al_2O_3

   $CsCl_2$ is the incorrect formula. Cs forms a 1+ cation while Cl forms a 1- anion. The correct formula for cesium chloride is CsCl.

7. Which compound involves *both* ionic and covalent bonding?
   Cl_2, Na_2SO_4, KCl, HF, HCN

Ionic compounds form between a metal and a nonmetal OR a metal and a polyatomic ion. Covalent compounds form between two or more nonmetals. In this list, Na_2SO_4 is the only molecule containing both bonding types. The bond between the two Na^+ ions and the SO_4^{2-} polyatomic anion is ionic in nature. However, the bonds forming the SO_4^{2-} polyatomic anion molecule are covalent since S and O are both nonmetals.

8. Which molecule *does not* have a dipole moment (which is nonpolar?)
   BrCl, ClF, BrF, O_2, ICl

   O_2 does not contain a dipole. Dipoles result from the unequal sharing of an electron pair in covalent bonds. Unequal sharing is the result of two atoms in a bond having different electronegativities. Since O_2 contains O bound to O, there is no difference in electronegativity and no dipole.

Module 9
Chemical Bonding

Introduction

This module explains how chemical bonds are formed. There are two basic types of chemical bonds: ionic and covalent. This module's goals include:

1. determining if a compound is ionic or covalent
2. drawing Lewis dot structures of atoms
3. writing formulas of simple ionic compounds
4. determining relative ionic bond strengths
5. drawing Lewis dot structures of ionic and covalent compounds
6. recognizing if a covalent bond is polar or nonpolar.

A periodic table will help you understand the electron configurations used in this chapter.

Module 9 Key Equations & Concepts

$$\text{Force of attraction between 2 ions} \propto \frac{q^+ \times q^-}{d^2}$$

$$\text{Force of attraction between 2 ions} \propto \frac{(\text{Charge on cation})(\text{Charge on anion})}{(\text{Distance between the ions})^2}$$

Coulomb's Law describes the attractive force between two ions of opposite charge. It is used to determine ionic bond strengths.

Sample Exercise

Determining if a Compound is Ionic or Covalent
1. Indicate which compounds are ionic in nature and which are covalent in nature.
$$CH_4, \ KBr, \ Ca_3N_2, \ Cl_2O_7, \ H_2SO_4, \ InCl_3$$

The correct answer is: ionic = KBr, Ca_3N_2 and $InCl_3$
covalent = CH_4, Cl_2O_7, and H_2SO_4

Look for metallic elements! Ionic compounds are formed by the reaction of metallic elements with nonmetallic elements or the reaction of the ammonium ion, NH_4^+, with nonmetals. Covalent compounds are formed by the reaction of two or more nonmetals

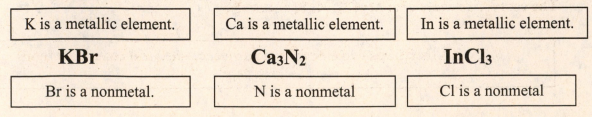

K is a metallic element.	Ca is a metallic element.	In is a metallic element.
KBr	**Ca₃N₂**	**InCl₃**
Br is a nonmetal.	N is a nonmetal	Cl is a nonmetal

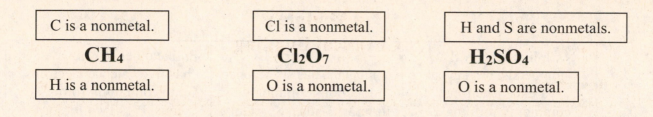

C is a nonmetal.	Cl is a nonmetal.	H and S are nonmetals.
CH₄	**Cl₂O₇**	**H₂SO₄**
H is a nonmetal.	O is a nonmetal.	O is a nonmetal.

Lewis Dot Structures of Atoms Sample Exercises
2. *Draw the correct Lewis dot structure of these elements: Mg, P, S, Ar*
   The correct Lewis dot structures are shown below.

 Use the periodic table to determine the element's number of valence electrons based upon its group number! The first step in drawing any Lewis dot structure *must* be determine how many valence electrons are present in the species.

The number next to each dot represents the order it was added to the structure. Essentially, each of the four sides of the element's symbol represents an orbital. One side represents an s orbital, and the remaining three are p orbitals. The s orbital is filled first, followed by the three p orbitals. Note that it does not matter where you start or whether you proceed clockwise or counterclockwise, as long as you follow Hund's rule and the Aufbau principle.

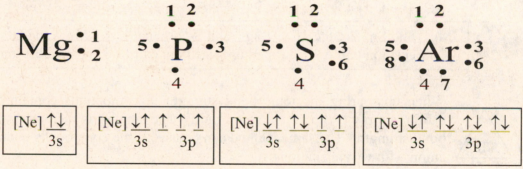

Lewis dot structures reflect the ground state elemental electronic structures including electron pairing. Notice that the orbital diagrams match the Lewis dot structures of each element.

Simple Ionic Compounds
3. *Write the correct formulas of the ionic compounds formed when Mg atoms react with: a) Cl atoms, b) S atoms, c) P atoms.*
   The correct answers are: MgCl₂, MgS, and Mg₃P₂

Mg, a IIA metal, has two electrons in its valence shell and commonly forms 2+ ions, Mg^{2+}.

MgCl₂	**MgS**	**Mg₃P₂**
Cl, like other VIIA nonmetals, has seven valence electrons commonly forming 1- ions, Cl⁻. Two Cl⁻ ions are required to balance the 2+ charge of the Mg forming neutral MgCl₂.	S, like other VIA nonmetals, has six valence electrons commonly forming 2- ions, S²⁻. Only one S²⁻ ion is required to neutralize the 2+ charge on the Mg²⁺ ion.	P, like other VA nonmetals, has five valence electrons and commonly forms 3- ions, P³⁻. Two P³⁻ ions balance the charge on three Mg²⁺ ions to form neutral Mg₃P₂.

4. *Arrange these ionic compounds in order of increasing ionic bond strength.*
MgSe, MgO, MgS
The correct answer is: MgSe < MgS < MgO

Coulomb's Law indicates that the force of attraction between ions $= \dfrac{q^+ \times q^-}{d^2}$. The strongest ionic bond has the largest charge and smallest ionic radii. Module 8 discusses ionic radii periodicity.

MgSe	**<**	**MgS**	**<**	**MgO**
Mg^{2+} and Se^{1-} are the largest ion pair.		Mg^{2+} and S^{2-} are the medium sized ion pair.		Mg^{2+} and O^{2-} are the smallest ion pair.

 TIP

When comparing the ionic bond strength in compounds containing a common cation, simply compare the anion radii. The smaller is the anion, the greater the bond strength. The same method can be used if the compounds contain a common anion with varying cations.

Drawing Ionic Compound Lewis Dot Structures
5. *Draw Lewis dot structures for each of these compounds.*
AlP, NaCl, MgCl₂
The correct structures are shown below.

When counting valence electrons, remember that a cation has, per positive charge, one electron less than the neutral parent atom. For each negative charge, anions have one electron more than the neutral parent atom.

$$Al^{3+} \left[: \ddot{P} : \right]^{3-} \qquad Na^{+} \left[: \ddot{\ddot{C}l} : \right]^{-} \qquad Mg^{2+} \, 2\left[: \ddot{\ddot{C}l} : \right]^{-}$$

Al loses all three of its valence electrons forming a 3+ ion. Thus, it has no dots.

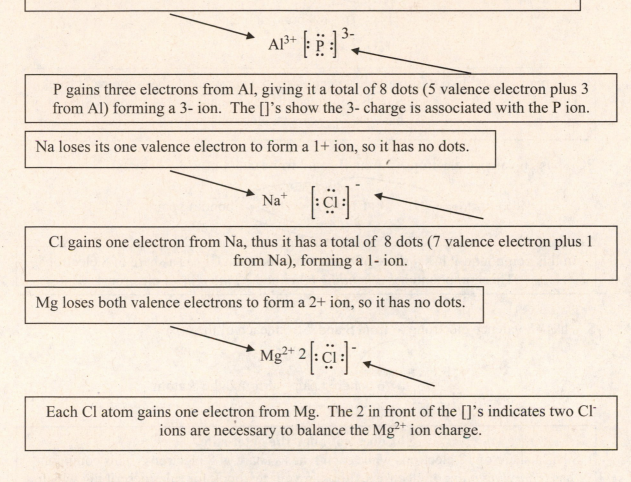

P gains three electrons from Al, giving it a total of 8 dots (5 valence electron plus 3 from Al) forming a 3- ion. The []'s show the 3- charge is associated with the P ion.

Na loses its one valence electron to form a 1+ ion, so it has no dots.

Cl gains one electron from Na, thus it has a total of 8 dots (7 valence electron plus 1 from Na), forming a 1- ion.

Mg loses both valence electrons to form a 2+ ion, so it has no dots.

Each Cl atom gains one electron from Mg. The 2 in front of the []'s indicates two Cl⁻ ions are necessary to balance the Mg²⁺ ion charge.

Drawing Simple Covalent Compound Lewis Dot Structures
6. Draw correct Lewis dot structures for each of these compounds.
$$SiH_4, PCl_3, SF_6$$
The correct structures are:

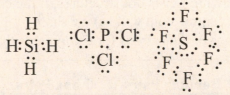

Try following these steps when drawing Lewis structures:
1. Determine the number of valence electrons in the compound.
2. Decide which atom is central atom then draw one bond (two electrons) to each of the remaining elements.
3. Fill in the octet for all elements, and count how many electrons were used. Procedures to apply when there are too many or too few electrons are discussed later in this module.

87

SiH₄ has 8 valence electrons (4 from Si and 1 from each of the 4 H)

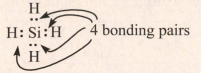

4 bonding pairs

In many cases, like this example, every atom in a compound obeys the octet rule. Thus, Si has a share of 8 electrons while each H has a share of 2 electrons. SiH₄ only has *bonding pairs* of electrons.

PCl₃ has 26 valence electrons (5 from P and 7 from each of the 3 Cl atoms).

lone pair

3 bonding pairs

In this compound P has a share of 8 electrons and each Cl has a share of 8 electrons. This compound has 3 *bonding pairs* and 1 *lone pair* of electrons.

SF₆ has 48 valence electrons (6 from S and 7 from each of the 6 F).

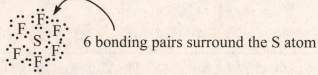

6 bonding pairs surround the S atom

SF₆ does not obey the octet rule.
S has a share of 12 electrons while each F has a share of 8 electrons. This compound has 6 *bonding pairs* of electrons. Look in your textbook for rules describing which compounds do not obey the octet rule.

INSIGHT: When drawing Lewis dot structures, if the compound obeys the octet rule, the central atom will have a share of 8 electrons. The possible combinations of 8 electrons for compounds that **obey the octet rule** are:

Bonding Pairs	Lone Pairs
4	0
3	1
2	2
1	3

INSIGHT: If the compound **does not obey the octet rule**, the central atom can have 2, 3, 5 or 6 pairs of electrons around the central atom in several bonding and lone pair combinations.

Compounds Containing Multiple Bonds

7. Draw correct Lewis dot structures for each of these compounds.

CO_2 *and* N_2

The correct structures are:

$$: \ddot{O} :: C :: \ddot{O} :$$

$$: N ::: N :$$

CO_2 has 16 valence electrons (4 from C and 6 from each of the two O atoms). C is the central atom. Connecting each O with the central C by one bonding pair and filling in all octets results in the following structure:

$$: \ddot{O} : \ddot{C} : \ddot{O} :$$

Note that this structure contains 20 electrons, four more than possible.

 TIP

To decrease electron numbers in a Lewis structure, make a double or triple bond. *If a double bond is made,* remove *one lone pair of electrons from each atom involved in the double bond.*

| Create a double bond here removing the circled lone pairs. | This structure still has too many electrons so repeat the process on the other side of the C atom. | The last structure is correct. Each atom has a share of 8 electrons with a total of 16 electrons. |

N_2 has 10 valence electrons. Connecting the two atoms and filling each octet results in a structure with 14 electrons:

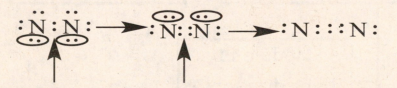

The formation of each multiple bond reduces the total electron count by 2 electrons. In this case, the process must be done twice to remove four electrons forming a *triple bond.*

89

Polar or Nonpolar Covalent Bonds in Compounds
8. Which compouns contains polar covalent bonds?
F_2, CH_4, H_2O

The correct answer is: CH_4 and H_2O contain polar covalent bonds and F_2 does not

The periodic trends regarding electronegativity are discussed in Module 8. Make sure you know those trends. You need that information for problems like this.

Polar covalent bonds occur when the two atoms involved in the bond have an electronegativity difference. In F_2 both atoms have the same electronegativity resulting in a nonpolar bond. In CH_4 and H_2O, the H to C or O bond involves atoms with different electronegativities. Consequently, there are polar covalent bonds in both CH_4 and H_2O.

> **INSIGHT:**
> Polar bonds have *dipoles* resulting from *partial positive* and *partial negative* charges on atoms from uneven electron sharing. We indicate dipoles by drawing an arrow over the bond with the head of the arrow pointing in the direction of the more electronegative atom. Each C-O bond in CO_2 is polar, as indicated below:
>
> :Ö :: C :: Ö:

Module 9 relates to some following Modules as shown in the graphic below.

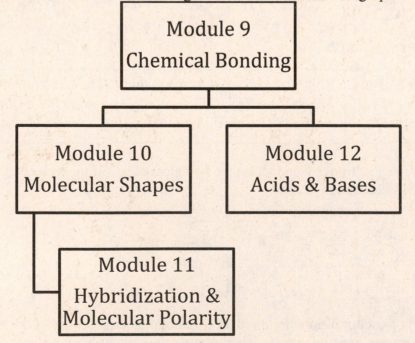

90

Module 10 Predictor Questions

The following questions may help you determine the extent you need to study this module. Questions are ranked according to ability.

      Level 1 = basic proficiency
      Level 2 = mid level proficiency
      Level 3 = high proficiency

If you can correctly answer Level 3 questions you probably do not need to spend much time on this module. If you can only answer Level 1 problems, you should review this module.

Level 1   1. Determine the electronic geometry and molecular shape of each molecule from its Lewis structure.
           a) CBr_4
           b) F_2
           c) SF_6
           d) PCl_5
           e) BF_3

Level 2   2. What is the molecular shape of PCl_3?

Level 3   3. Which species is *incorrectly* matched with its bond angles?

Molecule	Bond Angles
HCN	180°
ClO_3^-	slightly < 109°
NH_3	107°
SeO_4^{2-}	109.5°
CCl_4	90°, 120°, and 180°

Module 10 Predictor Question Solutions

1. Determine the electronic geometry and molecular shape of each molecule from its Lewis structure.

   a) CBr_4
   b) F_2
   c) SF_6
   d) PCl_5
   e) BF_3

Example	Electronic Geometry	Molecular Shape
CBr_4	Tetrahedral	Tetrahedral
F_2	Linear	Linear
SF_6	Octahedral	Octahedral
PCl_5	Trigonal bipyramidal	Trigonal bipyramidal
BF_3	Trigonal planar	Trigonal planar

2. What is the molecular shape of PCl_3?

The Lewis structure for PCl_3 contains 26 valence electrons. P is the central atom with three single bonds to Cl atoms and one lone pair. Accordingly, the electronic geometry is tetrahedral, but the lone pair on P results in a trigonal pyramidal molecular shape.

3. Which species is *incorrectly* matched with its bond angles?

Molecule	Bond Angles
HCN	180°
ClO_3^-	slightly < 109°
NH_3	107°
SeO_4^{2-}	109.5°
CCl_4	90°, 120°, and 180°

HCN has a linear electronic geometry and molecular shape resulting in 180° bond angles.

ClO_3^- has a tetrahedral electronic geometry, but one lone pair on the central Cl atom results in a trigonal pyramidal shape and bond angles slightly less than the standard 109° for tetrahedral molecules.

NH_3 also has a tetrahedral electronic geometry but one lone pair on the central N atom results in a trigonal pyramidal shape with bond angles of 107°.

SeO_4^{2-} has tetrahedral electronic geometry and molecular shape. Its bond angles are 109.5°.

CCl_4 has a tetrahedral electronic and molecular shape. Its bond angles are 109.5° NOT 90°, 120°, and 180°.

Module 10
Molecular Shapes

Introduction
Molecular shape refers to the geometrical arrangement of atoms around the central atom in a molecule or polyatomic ion. This module will help you understand and predict the stereochemistry of some molecules. Molecular shapes are important in the chemical reactivity of numerous compounds. The most important goal of this module is to learn to:

1. predict and name electronic geometries and molecular shapes.

Module 10 Key Equations & Concepts

All of the molecules described in this module have two geometries that you know: the electronic geometry and molecular shape.

1) **Electronic geometry** considers all regions of high electron density including bonding pairs, lone pairs, and double or triple bonds.

2) **Molecular shape** only considers those electrons and atoms involved in bonding pairs or in double and triple bonds.

The molecular shape and electronic geometry differ in molecules having lone pairs of electrons on the central atom.

Regions of high electron density	Electronic geometry	Molecular shape	Bond angles	Examples
2	Linear	Linear	180°	BeF_2, BeH_2, $BeCl_2$
3	Trigonal planar	Trigonal planar	120°	BH_3, $AlCl_3$, BF_3
4	Tetrahedral	Tetrahedral, trigonal pyramidal, or bent	Vary	CH_4, SiH_4, PF_3, H_2O
5	Trigonal bipyramidal	Trigonal bipyramidal, see-saw, T-shaped, or linear	Vary	PF_5, SF_4, ClF_3, XeF_2
6	Octahedral	Octahedral, square pyramidal, or square planar	Vary	SF_6, IF_5, XeF_4

CAUTION Electronic geometries are based on the number of high electron density regions (bonding and nonbonding electron pairs) *around the central atom.* Bonds (single, double, or triple bonds) count as ONE region of electron density.

It is difficult to determine the correct electronic geometry or molecular shape if you are working with an incorrect Lewis structure!

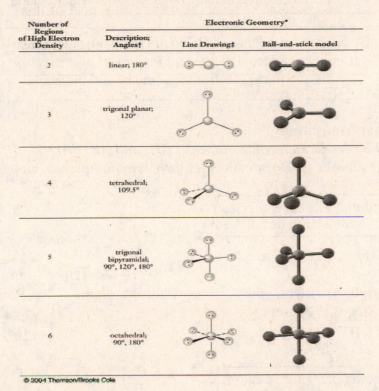

Number of Regions of High Electron Density	Description; Angles†	Electronic Geometry*	
		Line Drawing‡	Ball-and-stick model
2	linear; 180°		
3	trigonal planar; 120°		
4	tetrahedral; 109.5°		
5	trigonal bipyramidal; 90°, 120°, 180°		
6	octahedral; 90°, 180°		

© 2004 Thomson/Brooks Cole

Sample Exercises

Linear Molecules

1. What are the correct molecular shapes of BeI_2 and BeHF?
   The correct answer is: Both molecules have linear molecular shapes.

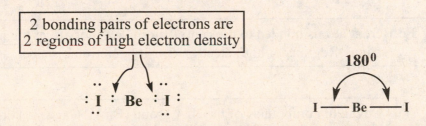

2 bonding pairs of electrons are
2 regions of high electron density

180^0

The linear shape is determined by the electrons surrounding the central Be atom, not the lone pairs on I.

95

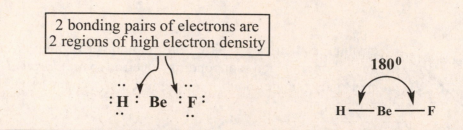

2 bonding pairs of electrons are
2 regions of high electron density

180^0

:H : Be : F:

H —— Be —— F

The linear shape is unaffected if there are two different atoms bonded to Be. The regions of high electron density determine molecular shape.

INSIGHT: Covalent Compounds of Be do not obey the octet rule. If Be is the central atom in a molecule there are 2 regions of high electron density giving a linear electronic geometry and molecular shape.

Trigonal Planar Molecules

2. What are the correct molecular shapes of BH₃ and AlHFBr?
  The correct answer is: Both molecules have trigonal planar molecular shapes.

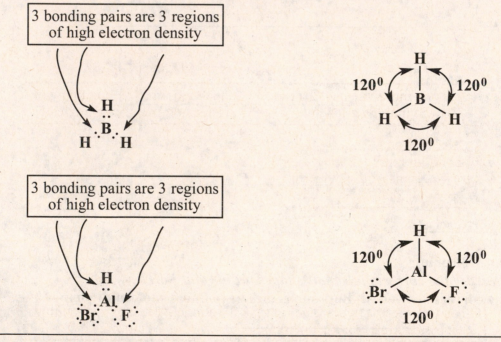

3 bonding pairs are 3 regions of high electron density

H
.B.
H   H

120^0 H 120^0
B
H   H
120^0

3 bonding pairs are 3 regions of high electron density

H
.Al.
.Br.   .F.

H
120^0 120^0
Al
.Br.   .F.
120^0

Having 3 different atoms bonded to the central atom does not affect the molecule's shape. Regions of high electron density determine shape.

INSIGHT: Covalent compounds of the IIIA group (B, Al, Ga, & In) do not obey the octet rule. If a IIIA element is the central atom, the molecule has 3 regions of high electron density around the central atom yielding a trigonal planar electronic geometry and molecular shape. This class of molecules has 3 bonding pairs of electrons.

Tetrahedral and Variations of Tetrahedral Molecules

3. What are the correct molecular shapes of SiH₄, PF₃, and H₂O?

The correct answers are: tetrahedral, trigonal pyramidal, and bent, respectively.

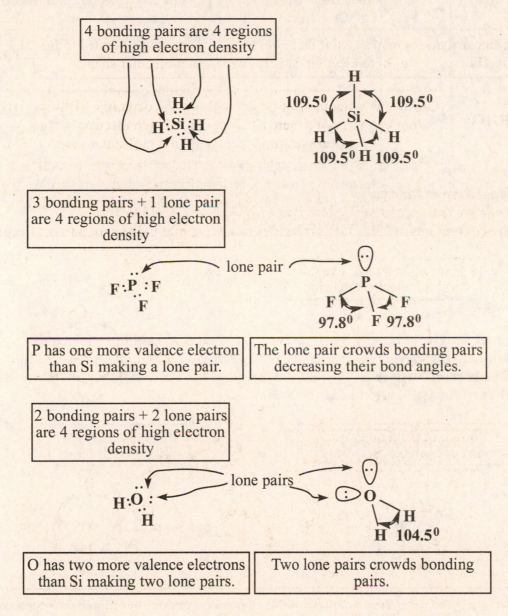

INSIGHT: Atoms in all molecules having four regions of high electron density obey the octet rule.

1. If a IVA element (C, Si, or Ge) is the central atom, electronic geometries and molecular shapes are tetrahedral. These molecules have 4 bonding pairs of electrons on central atom.
2. If a VA element (N, P, or As) is the central atom, the electronic geometry is tetrahedral and molecular shape is trigonal pyramidal. These molecules have 3 bonding pairs and 1 lone pair of electrons on central atom.
3. If a VIA element (O, S, Se) is the central atom, the electronic geometry is tetrahedral and the molecular shape is bent, angular, or V-shaped. These molecules have 2 bonding pairs and 2 lone pairs of electrons on central atom.
4. If a VIIA element (F, Cl, Br, or I) is the central atom, the electronic geometry is tetrahedral and the molecular shape is linear. These molecules have 1 bonding pairs and 3 lone pairs of electrons on central atom.

TIP — If the central atom has no lone pairs, the molecular shape and electronic geometry are the same.

Trigonal Bipyramidal and Variations of Trigonal Bipyramidal Molecules

4. What are the correct molecular shapes of PF$_5$, SF$_4$, ClF$_3$, and XeF$_2$

The correct answer is trigonal bipyramidal for PF$_5$, see-saw shaped for SF$_4$, T-shaped for ClF$_3$, and linear for XeF$_2$.

5 bonding pairs on P are 5 regions of high electron density.	PF$_5$ has a trigonal bipyramidal shape.

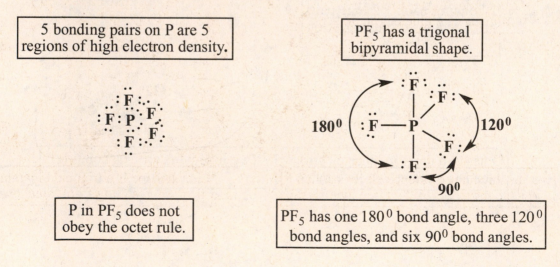

P in PF$_5$ does not obey the octet rule.	PF$_5$ has one 180^0 bond angle, three 120^0 bond angles, and six 90^0 bond angles.

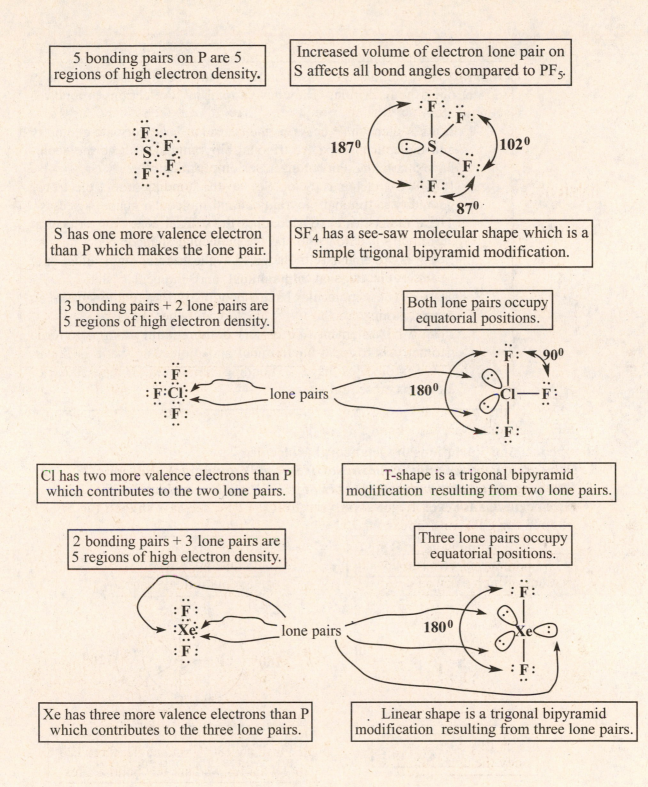

5 bonding pairs on P are 5 regions of high electron density.

Increased volume of electron lone pair on S affects all bond angles compared to PF_5.

S has one more valence electron than P which makes the lone pair.

SF_4 has a see-saw molecular shape which is a simple trigonal bipyramid modification.

3 bonding pairs + 2 lone pairs are 5 regions of high electron density.

Both lone pairs occupy equatorial positions.

Cl has two more valence electrons than P which contributes to the two lone pairs.

T-shape is a trigonal bipyramid modification resulting from two lone pairs.

2 bonding pairs + 3 lone pairs are 5 regions of high electron density.

Three lone pairs occupy equatorial positions.

Xe has three more valence electrons than P which contributes to the three lone pairs.

Linear shape is a trigonal bipyramid modification resulting from three lone pairs.

99

Central atoms in molecules based on five regions of high electron density do not obey the octet rule. They have a total of 10 electrons around the central atom.

1. If a VA element (P or As) is the central atom, electronic geometry and molecular shape are trigonal bipyramidal. These molecules have 5 bonding pairs of electrons on central atom.

2. If a VIA element (S or Se) is the central atom, electronic geometry is trigonal bipyramidal and molecular shape is seesaw. These molecules have 4 bonding pairs and 1 lone pair on central atom.

3. If a VIIA element (Cl, Br, or I) is the central atom, electronic geometry is trigonal bipyramidal and molecular shape is T-shaped. These molecules have 3 bonding pairs and 2 lone pairs on central atom.

4. If an VIIIA element (Xe or Kr) is the central atom, electronic geometry is trigonal bipyramidal and molecular shape is linear. These molecules have 2 bonding pairs and 3 lone pairs of electrons on central atom.

Octahedral and Variations of Octahedral Molecules

5. *What are the correct molecular shapes of SF_6, IF_5, and XeF_4?*

The correct answers are: octahedral for SF_6, square pyramidal for IF_5, and square planar for XeF_4.

6 bonding pairs on S are 6 regions of electron density.

SF_6 has an octahedral shape.

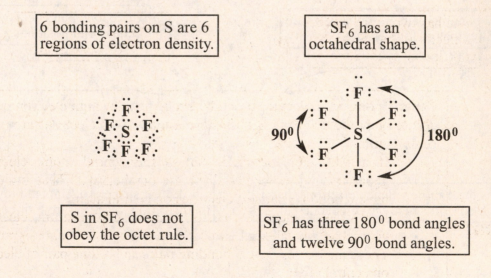

S in SF_6 does not obey the octet rule.

SF_6 has three 180^0 bond angles and twelve 90^0 bond angles.

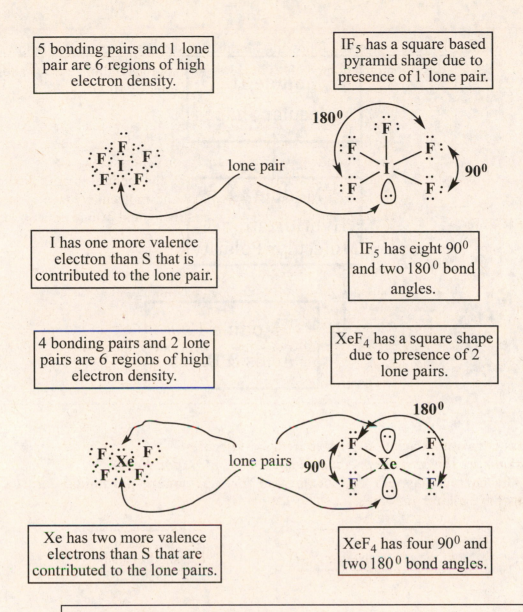

5 bonding pairs and 1 lone pair are 6 regions of high electron density.

IF_5 has a square based pyramid shape due to presence of 1 lone pair.

180⁰

lone pair

90⁰

I has one more valence electron than S that is contributed to the lone pair.

IF_5 has eight 90⁰ and two 180⁰ bond angles.

4 bonding pairs and 2 lone pairs are 6 regions of high electron density.

XeF_4 has a square shape due to presence of 2 lone pairs.

180⁰

lone pairs   90⁰

Xe has two more valence electrons than S that are contributed to the lone pairs.

XeF_4 has four 90⁰ and two 180⁰ bond angles.

INSIGHT: Central atoms in molecules based on six regions of high electron density do not obey the octet rule. They have a total of 12 electrons around the central atom.

1. If a VIA element (S or Se) is the central atom, electronic geometry and molecular shape are octahedral. These molecules have 6 bonding pairs of electrons on central atom.

2. If a VIIA element (Cl, Br, or I) is the central atom, electronic geometry is octahedral and molecular shape is square pyramidal. These molecules have 5 bonding pairs and 1 lone pair of electrons on central atom.

3. If an VIIIA element (Xe or Kr) is the central atom, electronic geometry is octahedral and molecular shape is square planar. These molecules have 4 bonding pairs and 2 lone pairs of electrons on central atom.

Module 10 relates to some following Modules as shown in the graphic below.

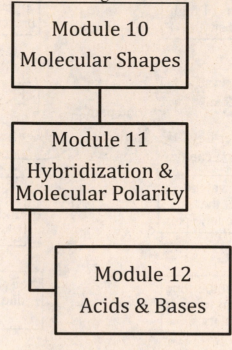

Module 10
Molecular Shapes

Module 11
Hybridization &
Molecular Polarity

Module 12
Acids & Bases

Module 11 Predictor Questions

The following questions may help you determine the extent you need to study this module. Questions are ranked according to ability.

       Level 1 = basic proficiency

       Level 2 = mid level proficiency

       Level 3 = high proficiency

If you can correctly answer Level 3 questions you probably do not need to spend much time on this module. If you can only answer Level 1 problems, you should review this module.

Level 1   1. What kind of hybrid orbitals are utilized by the carbon atom in a CF_4 molecule?

Level 1   2. According to valence bond theory, what is the hybridization of the sulfur atom in SF_6?

Level 2   3. What is the hybridization of a carbon atom involved in a triple bond?

Level 2   4. Show the bond dipoles for the following molecules.
CO_2, NI_3, OF_2, CH_2Cl_2

Level 1   5. Determine which molecule is nonpolar.
CCl_4, CH_2Cl_2, CH_3Cl, $CHCl_3$, SiH_2Cl_2

Level 3   6. Which is a *nonpolar* molecule with *polar* covalent bonds?
NH_3, H_2Te, $SOCl_2$ (S is the central atom), $BeBr_2$, HF

Module 11 Predictor Question Solutions

1. What kind of hybrid orbitals are utilized by the carbon atom in a CF_4 molecule?

    CF_4 is a tetrahedral molecule containing four C-F single bonds. There are four electron groups surrounding the central C atom which corresponds to sp^3 hybridization.

2. According to valence bond theory, what is the hybridization of the sulfur atom in SF_6?

    SF_6 is an octahedral molecule containing six S-F single bonds. There are six electron groups surrounding the central S atom which corresponds to sp^3d^2 hybridization.

3. What is the hybridization of a carbon atom involved in a triple bond?

    Carbon forms four bonds. A triply bonded C atoms has two regions of high electron density (one from the triple bond, one from the remaining single bond). This corresponds to sp hybridization.

4. Show the bond dipoles for the following molecules.
    CO_2, NI_3, OF_2, CH_2Cl_2

5. Determine which molecule is nonpolar.
    CCl_4, CH_2Cl_2, CH_3Cl, $CHCl_3$, SiH_2Cl_2

    In symmetrical molecules, bond dipoles can cancel one another. The nonpolar molecule in this list is CCl_4.

6. Which is a *nonpolar* molecule with *polar* covalent bonds?
    NH_3, H_2Te, $SOCl_2$ (S is the central atom), $BeBr_2$, HF

    In polar molecules bond dipoles do not cancel. All of the molecules in this list contain polar bonds but only in $BeBr_2$, having linear geometry, do the dipoles cancel resulting in a nonpolar molecule.

Module 11
Hybridization and Molecular Polarity

Introduction

Valence Bond theory is another way to describe molecular shapes. Valence bond theory involves hybridization (mixing) of atomic orbitals. Hybrid orbital names are derived from the orbitals used to make the hybrid. This module will help you understand and predict hybridization of atoms in molecules. Polarity refers to whether molecular electron density is symmetrical or asymmetrical. The goals of this module are:

1. learn to predict the hybridization of atoms using Valence Bond theory
2. understand the hybridization and geometry of double and triple bonds
3. learn how to determine molecular polarity.

Module 11 Key Equations & Concepts

1. **sp hybridized atoms**
   Atoms having *two regions of electron density* and a *linear electronic geometry* have hybrid orbitals made from one s and one p orbital.

2. **sp^2 hybridized atoms**
   Atoms having *three regions of electron density* and a *trigonal planar electronic geometry* have hybrid orbitals made from one s and two p orbitals.

3. **sp^3 hybridized atoms**
   Atoms having *four regions of electron density* and a *tetrahedral electronic geometry* have hybrid orbitals made from one s and three p orbitals.

4. **sp^3d hybridized atoms**
   Atoms having *five regions of electron density* and a *trigonal bipyramidal electronic geometry* have hybrid orbitals made from one s, three p, and one d orbitals.

5. **sp^3d^2 hybridized atoms**
   Atoms having *six regions of electron density* and an *octahedral electronic geometry* have hybrid orbitals made from one s, three p, and two d orbitals.

The number of regions of electron density describes both the electronic geometry and hybridization.

Regions of Electron Density	Electronic Geometry	Hybridization
2	Linear	sp
3	Trigonal planar	sp^2
4	Tetrahedral	sp^3
5	Trigonal bipyramidal	sp^3d
6	Octahedral	sp^3d^2

TIP

Counting regions of high electron density surrounding an atom is the key to questions involving geometry and hybridization in doubly and triply bonded compounds. Double and triple bonds count as one electron density region. Lone pairs are also one electron density region.

INSIGHT:

> The number of hybrid orbitals formed is equal to the number of atomic orbitals combined. For example:
> one s orbital + two p orbitals = three sp^2 hybrid orbitals

Sample Exercises
Hybridization
1. What is the hybridization of the underlined atom in these molecules?
$$\underline{Be}I_2, \ \underline{B}H_3, \ \underline{Si}H_4, \ \underline{P}F_5, \ \underline{S}F_6$$

The correct answers are: sp, sp^2, sp^3, sp^3d, and sp^3d^2, respectively.

INSIGHT:

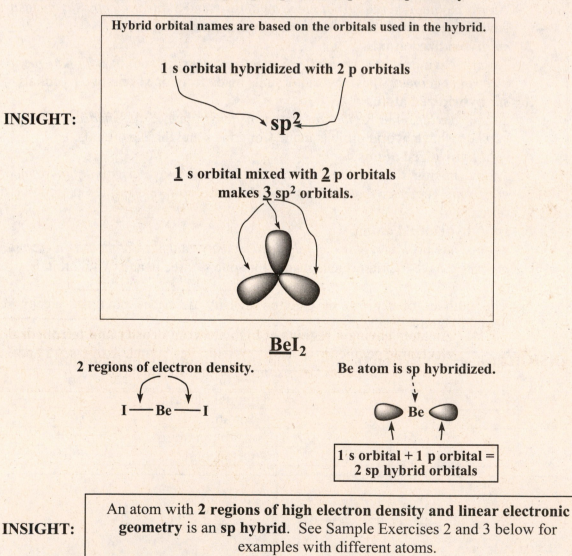

Hybrid orbital names are based on the orbitals used in the hybrid.

1 s orbital hybridized with 2 p orbitals

sp^2

1 s orbital mixed with **2** p orbitals
makes **3** sp^2 orbitals.

$\underline{Be}I_2$

2 regions of electron density.

I —— Be —— I

Be atom is sp hybridized.

Be

1 s orbital + 1 p orbital =
2 sp hybrid orbitals

INSIGHT:

> An atom with **2 regions of high electron density and linear electronic geometry** is an **sp hybrid**. See Sample Exercises 2 and 3 below for examples with different atoms.

106

$\underline{B}H_3$

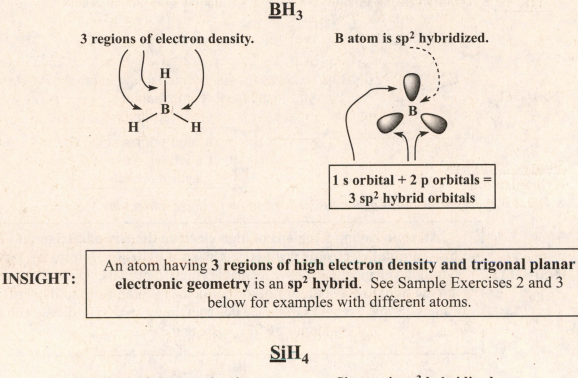

3 regions of electron density.

B atom is sp² hybridized.

1 s orbital + 2 p orbitals =
3 sp² hybrid orbitals

INSIGHT: An atom having **3 regions of high electron density and trigonal planar electronic geometry** is an **sp² hybrid**. See Sample Exercises 2 and 3 below for examples with different atoms.

$\underline{Si}H_4$

4 regions of electron density.

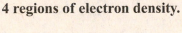

Si atom is sp³ hybridized.

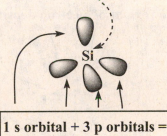

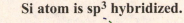

1 s orbital + 3 p orbitals =
4 sp³ hybrid orbitals

INSIGHT: An atom having **4 regions of high electron density and tetrahedral electronic geometry** is an **sp³ hybrid**. See Sample Exercises 2 and 3 below for examples with different atoms.

$\underline{P}F_5$

5 regions of electron density.

P atom is sp³d hybridized.

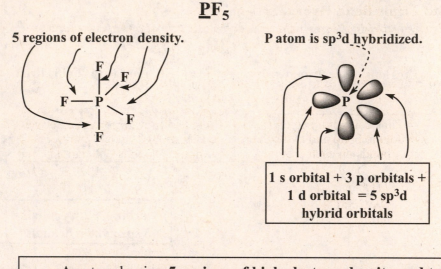

$$\boxed{\text{1 s orbital + 3 p orbitals +}\\ \text{1 d orbital }= 5\text{ sp}^3\text{d}\\ \text{hybrid orbitals}}$$

INSIGHT:

An atom having **5 regions of high electron density and trigonal bipyramidal electronic geometry** is an **sp³d hybrid**. To form an sp³d hybrid the element must have available empty d orbitals. Only elements on the third to the sixth row of the periodic table can form sp³d hybrid orbitals. Some examples of molecules containing sp³d hybridized central atoms are SF_4, ClF_3, and XeF_2.

$\underline{S}F_6$

6 regions of electron density.

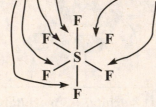

S atom is sp³d² hybridized.

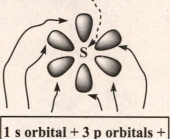

$$\boxed{\text{1 s orbital + 3 p orbitals +}\\ \text{2 d orbitsl }= 6\text{ sp}^3\text{d}^2\\ \text{hybrid orbitals}}$$

INSIGHT:

An atom having **6 regions of high electron density and octahedral electronic geometry** is an **sp³d² hybrid**. Just as for sp³d hybrids, there must be empty d orbitals available to form an sp³d² hybrid. Examples of molecules containing sp³d² hybridized atoms are BrF_5, and XeF_4.

Double and Triple Bond Hybridization
2. *What is the hybridization of the underlined atoms in these molecules?*
$\underline{C}_2H_4$, $\underline{C}_2H_2$, $H_2C\underline{O}$

The correct answers are: sp^2, sp, and sp^2, respectively.

C_2H_4

This compound contains a double bond and two single bonds on each C atom.

There are three regions of electron density around each C atom. The double bond is one region.

The single bonds are regions two and three.

Just as for BH_3, there are three regions of high electron density surrounding the C atom and the atom is sp^2 hybridized.

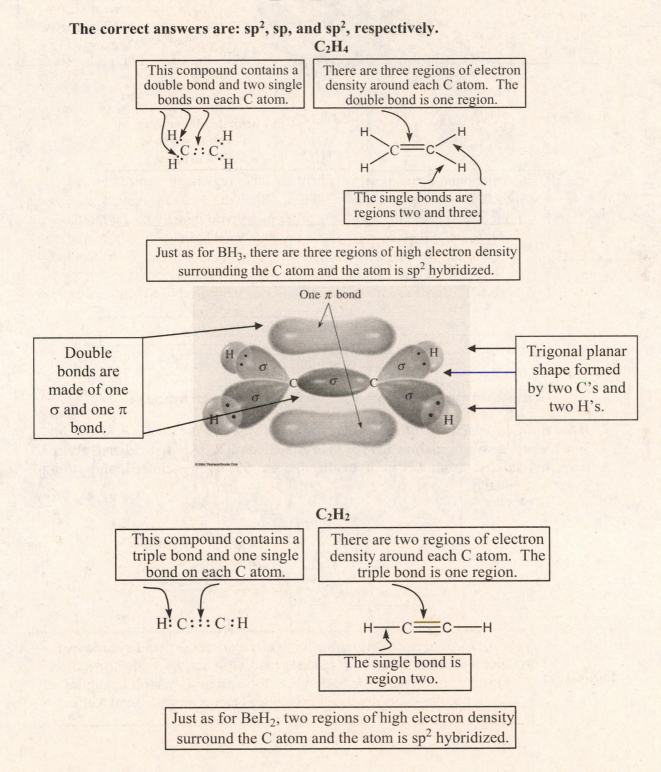

One π bond

Double bonds are made of one σ and one π bond.

Trigonal planar shape formed by two C's and two H's.

© 2004 Thomson/Brooks Cole

C_2H_2

This compound contains a triple bond and one single bond on each C atom.

There are two regions of electron density around each C atom. The triple bond is one region.

The single bond is region two.

Just as for BeH_2, two regions of high electron density surround the C atom and the atom is sp^2 hybridized.

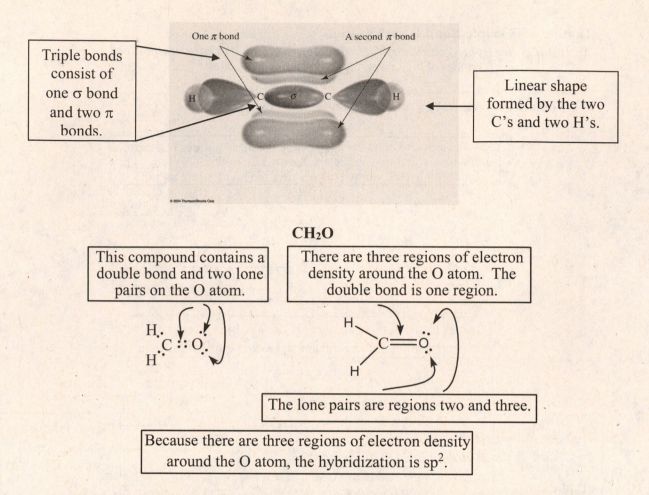

One π bond   A second π bond

Triple bonds consist of one σ bond and two π bonds.

H — C σ C — H

Linear shape formed by the two C's and two H's.

© 2004 Thomson/Brooks Cole

CH₂O

This compound contains a double bond and two lone pairs on the O atom.

There are three regions of electron density around the O atom. The double bond is one region.

H
 C ∷ O
H

H
  C = O
H

The lone pairs are regions two and three.

Because there are three regions of electron density around the O atom, the hybridization is sp².

3. *What is the hybridization of each of the indicated atoms in the amino acid alanine?*
   The correct answer is: atom 1 is sp³ hybridized, atom 2 is sp³ hybridized, atom 3 is sp³ hybridized, atom 4 is sp² hybridized, atom 5 is sp² hybridized, and atom 6 is sp³ hybridized

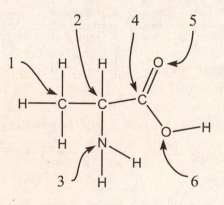

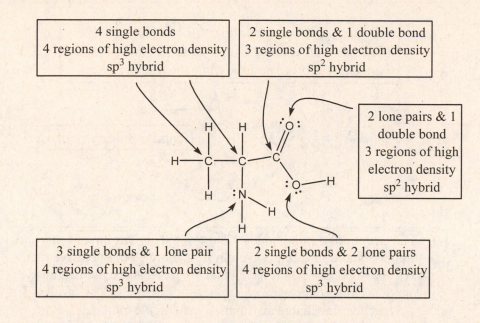

| 4 single bonds 4 regions of high electron density sp³ hybrid | 2 single bonds & 1 double bond 3 regions of high electron density sp² hybrid |

2 lone pairs & 1 double bond 3 regions of high electron density sp² hybrid

| 3 single bonds & 1 lone pair 4 regions of high electron density sp³ hybrid | 2 single bonds & 2 lone pairs 4 regions of high electron density sp³ hybrid |

INSIGHT: You must include the lone pairs to correctly answer this question. See Modules 9 and 10 for help in assigning lone pairs of electrons.

Molecular Polarity

4. Which of these molecules are polar?

BH_3, CH_2F_2, H_2O, SF_6

The correct answer is: BH_3 and SF_6 are nonpolar; CH_2F_2 and H_2O are polar.

Polar molecules have two essential features.
1) The molecule must contain at least one polar bond or one lone pair of electrons.   2) The molecule or its regions of high electron density must be asymmetrical so that the bond dipoles do not cancel each other.

Determining whether or not polar bonds cancel each other can be difficult. Imagine the central atom as a ball with strings attached to it correlating to the molecule's bonds. If you pull on each string simultaneously, will the ball move? If the answer is yes, then the molecule is polar. Keep in mind that you have to "pull" with different strengths if the atoms attached to the central atom are not all the same.

BH$_3$ contains polar bonds but the B-H bonds are symmetrical. Thus the dipoles for the polar bonds cancel each other and the molecule is nonpolar.

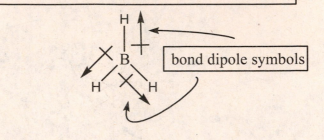

CH$_2$F$_2$ contains 4 polar bonds. The two C-H bonds have their bond dipole pointed toward the C atom (C is more electronegative than H). The two C-F bonds have bond dipoles that are pointed away from the C atom (F is more electronegative than C). The result is an asymmetrical charge distribution making the molecule polar.

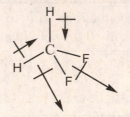

INSIGHT: Because CH$_2$F$_2$ is tetrahedral, every possible arrangement of the atoms is polar.

H$_2$O has two polar bonds and two lone pairs. Both bond dipoles for the O-H bonds are directed toward the O atom (O is more electronegative than H). These reinforce the large negative effect of the lone pairs making H$_2$O quite polar.

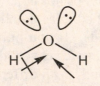

SF$_6$ has six polar S-F bonds. All boind dipoles are directed away from S toward F. This is a symmetrical arrangment of the bond dipoles leading to a nonpolar molecule.

Module 11 relates to some following Modules as shown in the graphic below.

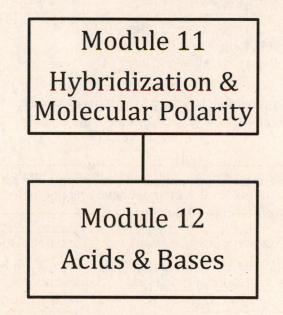

Practice Test Three
Modules 7-11

Level 3

1. Write the ground state electron configuration of Ni then answer the following questions.

   a) What is the principle quantum number, n, for the d electrons in Ni?
   b) How many d electrons in the Ni ground state are paired?
   c) How many d electrons in the Ni ground state are unpaired?
   d) What is the value of ℓ for the d electrons in Ni?

Level 3

2. How many electrons in a Au atom have the following quantum number combinations?

   a) $n = 3, \ell = 2$
   b) $n = 1, \ell = 0$
   c) $n = 5, \ell = 3, m_\ell = +2, m_s = -1/2$
   d) $n = 3, \ell = 1$
   e) $n = 4, \ell = 0, m_\ell = 0$

Level 2

3. From the list below select the most electronegative element and the element with the largest first ionization energy.
   Na, O, N, Al

Level 3

4. Select the element that releases the most energy when it accepts an electron. Does this element have a very positive or a very negative electron affinity value?
   Rb, Cs, I, Cl

Level 2

5. Explain why atomic radii increase down a group and from right to left across a period of the periodic table.

Levels 1-2

6. Draw the Lewis structure of MgO. Is MgO covalent or ionic? How many ions are present in one formula unit of MgO?

Levels 1-2

7. Draw the Lewis structure of SF_4. Is this molecule covalent or ionic? How many ions are present in one molecule of SF_4?

Levels 1-2

8. Determine both the electronic geometries and molecular shapes of these molecules or ions: XeF_4, I_3^-, CO_2, C_2H_4

Level 1

9. Which molecules or ions in question 8 are polar?

114

Level 1   10. Determine the orbital hybridization for the central atom of each
molecule in question 8 (there are two central atoms in C_2H_2).

Module 12 Predictor Questions

The following questions may help you determine the extent you need to study this module. Questions are ranked according to ability.

        Level 1 = basic proficiency
        Level 2 = mid level proficiency
        Level 3 = high proficiency

If you can correctly answer Level 3 questions you probably do not need to spend much time on this module. If you can only answer Level 1 problems, you should review this module.

Level 1
1. Identify the following as Arrhenius acids, Arrhenius bases, or neither.
   a) H_3BO_3
   b) RbOH
   c) $Ca(OH)_2$
   d) $C_2H_3O_2H$
   e) NaOH
   f) HNO_3

Level 1
2. According to Brønsted-Lowry theory, a base is defined as:
   a) an electron pair acceptor
   b) a proton acceptor
   c) an electron pair donor
   d) a proton donor
   e) any species that can produce hydroxide ions in aqueous solution

Level 1
3. According to Lewis theory, a base is defined as:
   a) a proton acceptor
   b) a proton donor
   c) an electron-pair donor
   d) any compound that contains electron pairs
   e) an electron-pair acceptor

Level 1
4. Which of these **cannot** be a Brønsted-Lowry acid?
   HCl, H_2O, CaO, NH_4^+, CH_3COOH

Level 2
5. Which one of these is a weak acid?
   HNO_3, H_3PO_4, $HClO_3$, $HClO_4$, HI

Level 3   6. In the following reaction,

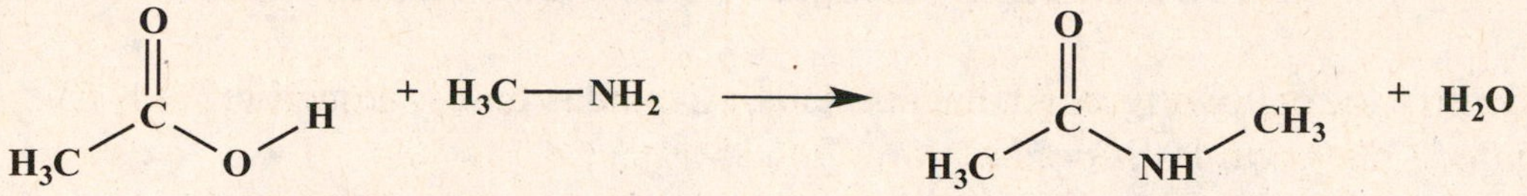

CH$_3$NH$_2$ reacts as:
a) only an Arrhenius base
b) only a Lewis base
c) only a Brønsted-Lowry base
d) a Brønsted-Lowry and Lewis base
e) an Arrhenius, Brønsted-Lowry, and Lewis base

Level 3   7. According to Brønsted-Lowry theory, which of these anions is the strongest base?
NO$_3^-$, Cl$^-$, CN$^-$, ClO$_4^-$, HSO$_4^-$

Level 2   8. Identify the conjugate acid-base pairs in the reactions below.
a) H$_2$S + NH$_3$ → HS$^-$ + NH$_4^+$
b) H$_2$O + SO$_3^{2-}$ → OH$^-$ + HSO$_3^-$
c) HF + H$_2$O → F$^-$ + H$_3$O$^+$

Level 3   9. Arrange these in order of increasing base strength.
HSO$_4^-$, HSeO$_4^-$

Module 12 Predictor Question Solutions

1. Identify the following as Arrhenius acids, Arrhenius bases, or neither.
   - a) H_3BO_3
   - b) RbOH
   - c) $Ca(OH)_2$
   - d) $C_2H_3O_2H$
   - e) NaOH
   - f) HNO_3

   Arrhenius acids are molecules containing hydrogen that ionize in water to produce H^+. Arrhenius bases contain hydroxide groups and dissociate in water to produce OH^- ions.

a) H_3BO_3	**Arrhenius acid**
b) RbOH	**Arrhenius base**
c) $Ca(OH)_2$	**Arrhenius base**
d) $C_2H_3O_2H$	**Arrhenius acid**
e) NaOH	**Arrhenius base**
f) HNO_3	**Arrhenius acid**

2. According to the Brønsted-Lowry theory, a base is defined as:
   - a) an electron pair acceptor
   - b) a proton acceptor
   - c) an electron pair donor
   - d) a proton donor
   - e) any species that can produce hydroxide ions in aqueous solution

   b) Brønsted-Lowry bases are proton acceptors.

3. According to Lewis theory, a base is defined as:
   - a) a proton acceptor
   - b) a proton donor
   - c) an electron-pair donor
   - d) any compound that contains electron pairs
   - e) an electron-pair acceptor

   c) Lewis bases are electron pair donors.

4. Which of these *cannot* be a Brønsted-Lowry acid?
   HCl, H_2O, CaO, NH_4^+, CH_3COOH
   CaO cannot act as a Bronsted-Lowry acid because it has no proton (H^+) to donate.

5. Which one of these is a weak acid?
   HNO_3, H_3PO_4, $HClO_3$, $HClO_4$, HI

H_3PO_4 is a weak acid. You should memorize the seven strong acids and remember that all other acids are weak.

6. In the following reaction,

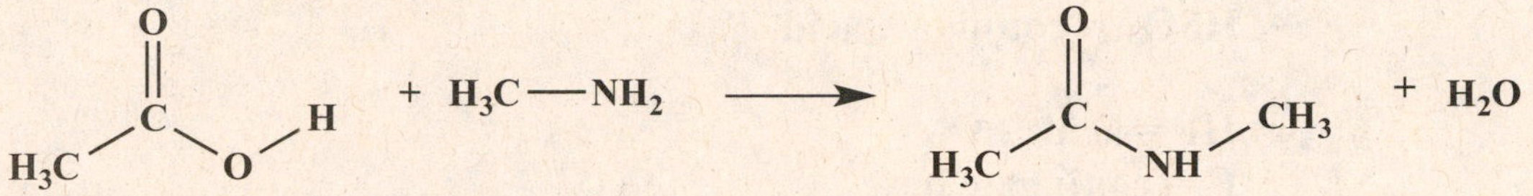

CH$_3$NH$_2$ reacts as:
        a) only an Arrhenius base
        b) only a Lewis base
        c) only a Brønsted-Lowry base
        d) a Brønsted-Lowry base and a Lewis base
        e) Arrhenius, Brønsted-Lowry, and Lewis bases

b) In forming the C-N bond, CH$_3$NH$_2$ donates an electron pair acting as a Lewis base.

7. According to Brønsted-Lowry theory, which of these anions is the strongest base?
   NO_3^-, Cl^-, CN^-, ClO_4^-, HSO_4^-

The easiest way to evaluate base strength is to study their conjugate acids. Weak acids have strong conjugate bases while strong acids have weak conjugate bases.

Base	Conjugate Acid	Acid Strength
NO_3^-	HNO_3	strong
Cl^-	HCl	strong
CN^-	HCN	weak
ClO_4^-	$HClO_4$	strong
HSO_4^-	H_2SO_4	strong

Since HCN is the only weak acid, its conjugate base (CN$^-$) is the strongest base in the list.

8. Identify the conjugate acid-base pairs in the reactions below.
        a) $H_2S + NH_3 \rightarrow HS^- + NH_4^+$
        b) $H_2O + SO_3^{2-} \rightarrow OH^- + HSO_3^-$
        c) $HF + H_2O \rightarrow F^- + H_3O^+$

Conjugate acid-base pairs differ only by the presence (acid) or absence (base) of a proton.
   a) H$_2$S = acid
      HS$^-$ = conjugate base

   NH$_3$ = base
   NH$_4^+$ = conjugate acid

b) H_2O = acid
   OH^- = conjugate base

   SO_3^{2-} = base
   HSO_3^- = conjugate acid

c) HF = acid
   F^- = conjugate base

   H_2O = base
   H_3O^+ = conjugate acid

9. Arrange these in order of increasing base strength.
   $$HSO_4^-, HSeO_4^-$$

Again, look at the conjugate acids to evaluate base strength. The conjugate acid of HSO_4^- is H_2SO_4, a strong acid. The conjugate acid of $HSeO_4^-$ is H_2SeO_4, a weak acid. Since H_2SeO_4 is a weak acid it has the stronger conjugate base.

$HSeO_4^-$ is a stronger base than HSO_4^-.

Module 12
Acids and Bases

Introduction

There are three common acid/base theories routinely discussed in general chemistry: Arrhenius, Brønsted-Lowry, and Lewis theories. This module will help you understand:

1. the distinctions and commonalities between the three theories
2. how to distinguish between compounds that behave as acids or bases in one theory but not in another.

Module 12 Key Concepts

1. **Arrhenius Acid-Base Theory definitions**

   Acid: produces protons (H^+) in aqueous solution

   Base: produces OH^- in aqueous solution

   Arrhenius is the most restrictive of the three theories since it requires both an aqueous solution and compounds having either an H^+ or OH^-.

2. **Brønsted-Lowry Acid-Base Theory definitions**

   Acid: proton donor

   Base: proton acceptor

   Brønsted-Lowry theory is less restrictive. Bases do not have to contain OH^- and the compounds do not have to be in aqueous solution.

3. **Lewis Acid-Base Theory definitions**

   Acid: electron pair donor

   Base: electron pair acceptor

   Lewis theory is the least restrictive of the theories, as it does not require the presence of protons, OH^-, or aqueous species.

Sample Exercise
Arrhenius Acid-Base Theory

1. Which of these compounds are Arrhenius acids and which are Arrhenius bases?
   HCl, $NaOH$, H_2SO_4, BCl_3, Na_2CO_3, $Ba(OH)_2$, C_2H_4

   The correct answer is: HCl and H_2SO_4 are Arrhenius acids; $NaOH$ and $Ba(OH)_2$ are Arrhenius bases. Na_2CO_3, BF_3 and C_2H_4 are neither Arrhenius acids nor bases.

To identify Arrhenius acids look for compounds that ionize in water forming H^+. To identify Arrhenius bases look for compounds that dissociate in water producing OH^-.

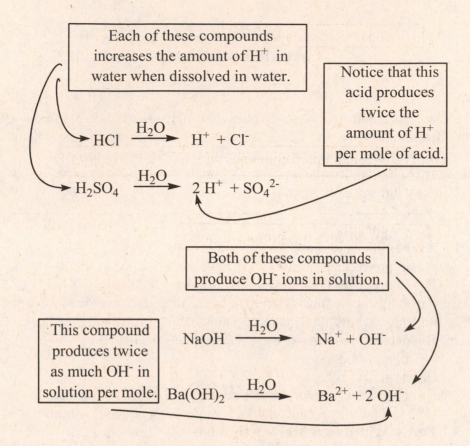

Each of these compounds increases the amount of H^+ in water when dissolved in water.

Notice that this acid produces twice the amount of H^+ per mole of acid.

$$HCl \xrightarrow{H_2O} H^+ + Cl^-$$

$$H_2SO_4 \xrightarrow{H_2O} 2\,H^+ + SO_4^{2-}$$

Both of these compounds produce OH^- ions in solution.

This compound produces twice as much OH^- in solution per mole.

$$NaOH \xrightarrow{H_2O} Na^+ + OH^-$$

$$Ba(OH)_2 \xrightarrow{H_2O} Ba^{2+} + 2\,OH^-$$

CAUTION It is relatively easy to see that BF_3 and Na_2CO_3 are not Arrhenius acids or bases since they do not contain H or OH. C_2H_4 is a little trickier. It contains H atoms. Do not let this confuse you! H atoms in C_2H_4 are not acidic because their C-H bond is too strong to be easily broken.

Brønsted-Lowry Acid-Base Theory
2. Which of these compounds can be classified as Brønsted-Lowry acids and bases?
HF, NH₃, H₂SO₄, BCl₃, Na₂CO₃, K₂S

The correct answer is: HF and H_2SO_4 are Brønsted-Lowry acids; NH_3 and Na_2CO_3 are Bronsted-Lowry bases. BCl_3 and K_2S are neither.

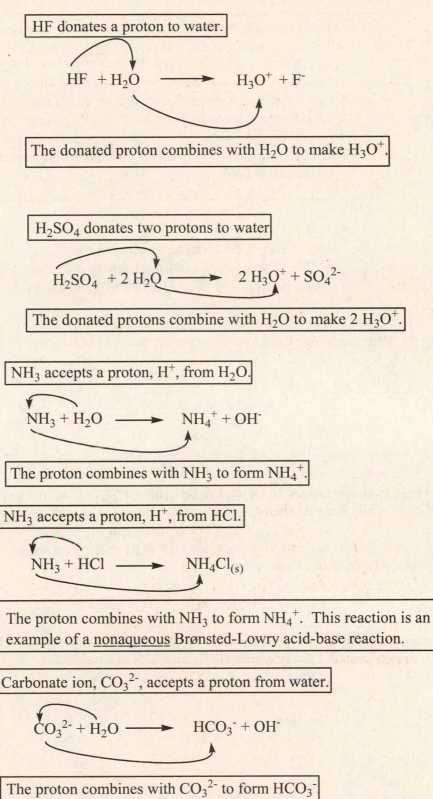

HF donates a proton to water.

$$HF + H_2O \longrightarrow H_3O^+ + F^-$$

The donated proton combines with H_2O to make H_3O^+.

H₂SO₄ donates two protons to water

$$H_2SO_4 + 2 H_2O \longrightarrow 2 H_3O^+ + SO_4^{2-}$$

The donated protons combine with H_2O to make 2 H_3O^+.

NH_3 accepts a proton, H^+, from H_2O.

$$NH_3 + H_2O \longrightarrow NH_4^+ + OH^-$$

The proton combines with NH_3 to form NH_4^+.

NH_3 accepts a proton, H^+, from HCl.

$$NH_3 + HCl \longrightarrow NH_4Cl_{(s)}$$

The proton combines with NH_3 to form NH_4^+. This reaction is an example of a <u>nonaqueous</u> Brønsted-Lowry acid-base reaction.

Carbonate ion, CO_3^{2-}, accepts a proton from water.

$$CO_3^{2-} + H_2O \longrightarrow HCO_3^- + OH^-$$

The proton combines with CO_3^{2-} to form HCO_3^-
The Na^+ ions are spectator ions in this reaction.

Anions of weak acids are Brønsted-Lowry bases. The carbonate ion, CO_3^{2-}, is the anion of the weak acid carbonic acid, H_2CO_3.

YIELD

- Because Arrhenius and Brønsted-Lowry acids are both proton donors, they are identified similarly.
- Brønsted-Lowry bases, however, may not contain hydroxide ions. Instead they accept a proton from water forming hydroxide ions in aqueous solutions.

Conjugate acid-base pairs: Based on Brønsted-Lowry theory, conjugate acid-base pairs differ by the presence or absence of a proton. Each Brønsted-Lowry acid has a conjugate base (having lost a proton) and each Brønsted-Lowry base has a conjugate acid.

3. Identify the Brønsted-Lowry acid-base conjugate pairs in these reactions.

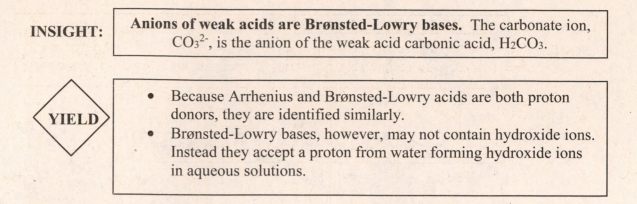

$$F^- \;+\; H_2O \;\rightleftharpoons\; HF \;+\; OH^-$$

The correct answers are: CH_3COOH is an acid; CH_3COO^- is its conjugate base
H_2O is a base; H_3O^+ is its conjugate acid
F^- is a base; HF is its conjugate acid
H_2O is an acid; OH^- is its conjugate base

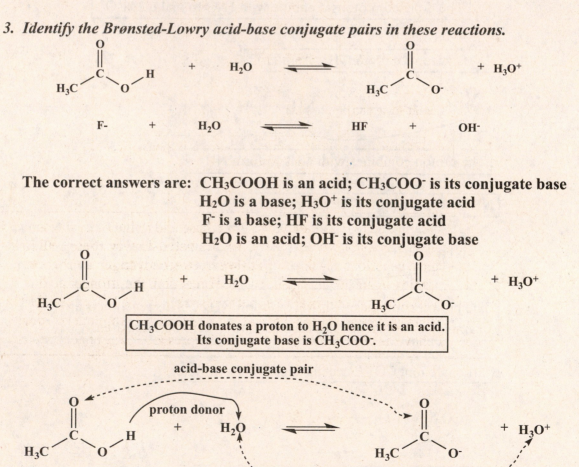

CH_3COOH donates a proton to H_2O hence it is an acid.
Its conjugate base is CH_3COO^-.

acid-base conjugate pair

proton donor

acid-base conjugate pair

H_2O accepts a proton from CH_3COOH hence it is a base.
Its conjugate acid is H_3O^+.

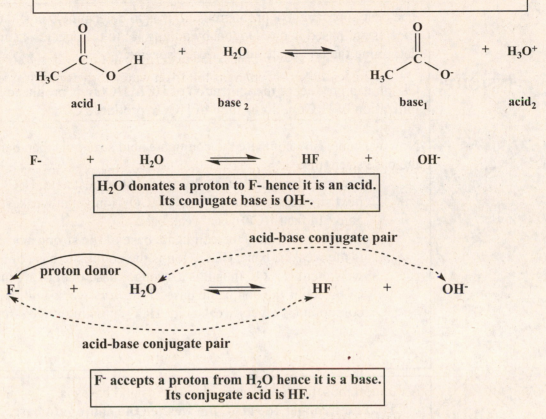

F^- + H_2O ⇌ HF + OH^-

H_2O donates a proton to F^- hence it is an acid.
Its conjugate base is OH^-.

acid-base conjugate pair

proton donor

F^- + H_2O ⇌ HF + OH^-

acid-base conjugate pair

F^- accepts a proton from H_2O hence it is a base.
Its conjugate acid is HF.

YIELD

Notice in one reaction above H_2O is a base and in the other it is an acid. Water is an *amphoteric* species. **In Brønsted-Lowry theory all acid-base reactions are a competition between stronger and weaker acids or bases.** In the CH_3COOH reaction, the stronger acid is CH_3COOH so water is a base in its presence. In the F^- reaction, H_2O is the stronger acid so it is the acid in this reaction. Amphoteric species behave as either an acid or base in the presence of stronger species.

125

4. Arrange the following species in order of increasing base strength.
HCO_3^-, Cl^-, CO_3^{2-}
The correct answer is: $Cl^- < HCO_3^- < CO_3^{2-}$

INSIGHT:

The best approach to a problem like this is to determine the acid that the species are based upon. Do this by adding H^+ to the species. Simply put determine the conjugate acids of each species.

Cl^- ion is a product of the ionization of HCl. HCO_3^- is produced by the ionization of H_2CO_3. Ionization of HCO_3^- produces CO_3^{2-}.

Now we can easily compare the conjugate acid strengths then determine the base strengths.
- For acid strengths, HCl is by far the strongest acid, H_2CO_3 is the next strongest acid, and finally HCO_3^- ion is the weakest acid. In fact, HCO_3^- is a weak base.
- Since the Cl^- ion is the conjugate base of the strong acid HCl, it is the weakest base. HCO_3^- ion is the conjugate base of the weak acid H_2CO_3, thus it is a stronger base than Cl^-. Finally, the CO_3^{2-} ion is the conjugate base of the very weak acid (a basic compound is a very weak acid) HCO_3^- making it the strongest base.

YIELD

An important set of mnemonics about acids and bases is:
1) **The stronger the acid, the weaker the conjugate base.**
2) **The weaker the acid, the stronger the conjugate base.**
3) **The stronger the base, the weaker the conjugate acid.**
4) **The weaker the base, the stronger the conjugate acid.**

Lewis Acid-Base Theory
5. Identify the Lewis acids and bases in the following reactions.
$$NH_3 + HCl \rightarrow NH_4Cl$$
$$BCl_3 + NH_3 \rightarrow BCl_3NH_3$$

The correct answer is: HCl and BCl_3 are Lewis acids; NH_3 is the Lewis base in both reactions.

YIELD

Lewis acids and bases are best determined using Lewis dot structures then watching the motion of the electron pairs. It is also helps to realize that Arrhenius and Brønsted-Lowry theories are focused on the protons, H^+. Lewis acid-base theory focuses on electrons. Consequently, the actions are reversed.

Acids are proton __donors__ in Arrhenius and Brønsted-Lowry theories.
In Lewis theory acids are electron pair __acceptors__.

126

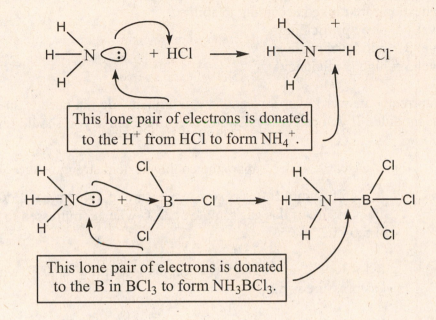

This lone pair of electrons is donated to the H^+ from HCl to form NH_4^+.

This lone pair of electrons is donated to the B in BCl_3 to form NH_3BCl_3.

Module 12 relates to some following Modules as shown in the graphic below.

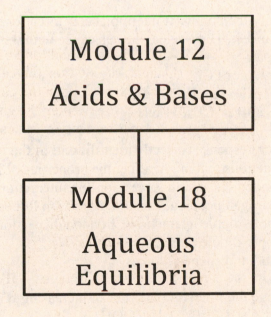

Module 12
Acids & Bases

Module 18
Aqueous
Equilibria

Module 13 Predictor Questions

The following questions may help you determine the extent you need to study this module. Questions are ranked according to ability.

    Level 1 = basic proficiency
    Level 2 = mid level proficiency
    Level 3 = high proficiency

If you can correctly answer Level 3 questions you probably do not need to spend much time on this module. If you can only answer Level 1 problems, you should review this module.

Level 1   1. A 6.25 L sample of gas exerts a pressure of 1.46 atm at 25°C. What is the pressure of this gas sample at 25°C when compressed to a volume of 5.05 L?

Level 2   2. If 1.47×10^{-3} mol of Ar gas occupies 75.0 mL at 26.0°C, what is the gas pressure in atm?

Level 1   3. What is the mass (in grams) of 207 mL of chlorine trifluoride gas at 0.920 atm and 45.0°C?

Level 3   4. Identify the dominant intermolecular force for each of these substances. Then, select the substance with the higher boiling point in each pair.
    a) $MgCl_2$ or PCl_3
    b) CH_3NH_2 or CH_3F
    c) CH_3OH or CH_3CH_2OH

Level 3   5. Which response correctly identifies all of the intermolecular interactions that might affect the properties of BrI?
    a) London dispersion forces, ion-ion interaction
    b) hydrogen bonding, London dispersion forces
    c) dipole-dipole interactions, London dispersion forces
    d) dipole-dipole interactions
    e) London dispersion forces

Level 2   6. Arrange the following in order of increasing boiling point.
    CaO, CCl_4, CH_2Br_2, CH_3COOH

Level 1   7. In a face-centered cubic lattice, how many atoms are present in one unit cell?

Level 2   8. A solid has a density of 5.42 g/cm^3 and crystallizes in a cubic unit cell with an edge length of 4.46×10^{-8} cm. If the substance has a molar mass of 144.24 g/mol, how many atoms are in one cell? Identify the cubic unit cell type.

Level 1   9. Under comparable conditions, how much faster will a sample of He effuse through a small opening than a sample of Cl_2?

Module 13 Predictor Question Solutions

1. A 6.25 L sample of gas exerts a pressure of 1.46 atm at 25°C. What is the pressure of this gas sample at 25°C when compressed to a volume of 5.05 L?

   $P_1V_1 = P_2V_2$

   $(1.46 \text{ atm})(6.25 \text{ L}) = P_2(5.05 \text{ L})$

   $P_2 = 1.81 \text{ atm}$

2. If 1.47×10^{-3} mol of Ar gas occupies 75.0 mL at 26.0°C, what is the gas pressure in atm?

   $T = 26°C + 273.15 = 299 \text{ K}$

   $V = 75.0 \text{ mL} = 0.0750 \text{ L}$

   $$PV = nRT \Rightarrow P = \frac{nRT}{V} = \frac{(1.47 \times 10^{-3} \text{ mol})(0.0821 \frac{L \cdot atm}{mol \cdot K})(299 \text{ K})}{0.0750 \text{ L}} = 0.481 \text{ atm}$$

   $T = 26°C + 273.15 = 299 \text{ K}$

   $V = 75.0 \text{ mL} = 0.0750 \text{ L}$

   $$PV = nRT \Rightarrow P = \frac{nRT}{V} = \frac{(1.47 \times 10^{-3} \text{ mol})(0.0806 \frac{L \cdot atm}{mol \cdot K})(299 \text{ K})}{0.0750 \text{ L}} = 0.472 \text{ atm}$$

3. What is the mass (in grams) of 207 mL of chlorine trifluoride gas at 0.920 atm and 45.0°C?

$$T = 45°C + 273.15 = 318 \text{ K}$$

$$V = 207 \text{ mL} = 0.207 \text{ L}$$

$$PV = nRT \Rightarrow n = \frac{PV}{RT} = \frac{(0.920 \text{ atm})(0.207 \text{ L})}{(0.0821 \frac{\text{L} \cdot \text{atm}}{\text{mol} \cdot \text{K}})(318 \text{ K})} = 7.30 \times 10^{-3} \text{ mol ClF}_3$$

$$(7.30 \times 10^{-3} \text{ mol ClF}_3)\left(\frac{92.45 \text{ g ClF}_3}{1 \text{ mol ClF}_3}\right) = 0.675 \text{ g ClF}_3$$

$$T = 45°C + 273.15 = 318 \text{ K}$$

$$V = 207 \text{ mL} = 0.207 \text{ L}$$

$$PV = nRT \Rightarrow n = \frac{PV}{RT} = \frac{(0.920 \text{ atm})(0.207 \text{ L})}{(0.08205 \frac{\text{L} \cdot \text{atm}}{\text{mol} \cdot \text{K}})(318 \text{ K})} = 7.30 \times 10^{-3} \text{ mol ClF}_3$$

$$(7.30 \times 10^{-3} \text{ mol ClF}_3)\left(\frac{92.45 \text{ g ClF}_3}{1 \text{ mol ClF}_3}\right) = 0.675 \text{ g ClF}_3$$

4. Identify the dominant intermolecular force for each of these substances. Then, select the substance with the higher boiling point in each pair.

> a) $MgCl_2$ or PCl_3
> b) CH_3NH_2 or CH_3F
> c) CH_3OH or CH_3CH_2OH
> d) C_6H_{14} or C_6H_{12}

a) $MgCl_2$ = ion-ion attraction
PCl_3 = dipole-dipole attraction

    Ion-ion attraction is the stronger force. $MgCl_2$ has a boiling point of 1412°C. PCl_3 has a boiling point of 76.1°C.

b) CH_3NH_2 = hydrogen bonding
CH_3F = dipole-dipole attraction

    Hydrogen bonding is the stronger force. CH_3NH_2 has a boiling point of -6°C. CH_3F has a boiling point of -78.4°C.

c) CH_3OH = hydrogen bonding
CH_3CH_2OH = hydrogen bonding

    These two molecules are very similar in structure and intermolecular force strength. In cases like this the molecule with the greater molecular weight has the higher boiling point. CH_3OH has a boiling point of 64.7°C. CH_3CH_2OH has a boiling point of 78.4°C

5. Which response correctly identifies all of the intermolecular interactions that will affect the properties of BrI?
   a) London dispersion forces, ion-ion interaction
   b) hydrogen bonding, London dispersion forces
   c) dipole-dipole interactions, London dispersion forces
   d) dipole-dipole interactions
   e) London dispersion forces

   The correct answer is c). BrI is a covalent, polar molecule with no possibility of forming hydrogen bonds. Its primary intermolecular force is dipole-dipole interactions. Additionally, all molecules have London dispersion forces.

6. Arrange the following in order of increasing boiling point.
   CaO, CCl_4, CH_2Br_2, CH_3COOH

   Higher boiling points are the result of stronger intermolecular forces. Intermolecular force strength has this relative ranking: ion-ion attraction > hydrogen bonding > dipole-dipole attractions > London dispersion forces.

   CaO = ion-ion attraction
   CCl_4 = London dispersion forces
   CH_2Br_2 = dipole-dipole attraction
   CH_3COOH = hydrogen bonding
   Boiling points:
   CCl_4 (76.7°C) < CH_2Br_2 (97.0°C) < CH_3COOH (118°C) < CaO (2850°C)

7. In a face-centered cubic lattice, how many atoms are present in one unit cell?

There are _four_ atoms per unit cell in a face centered cubic lattice.

8. A solid has a density of 5.42 g/cm³ and crystallizes in a cubic unit cell with an edge length of 4.46 x 10⁻⁸ cm. If the substance has a molar mass of 144.24 g/mol, how many atoms are in one cell? Identify the cubic unit cell type.

$$V = l^3 = (4.46 \times 10^{-8} \text{ cm})^3 = 8.87 \times 10^{-23} \text{ cm}^3$$

$$D = \frac{m}{V} \Rightarrow m = DV = (5.42 \frac{g}{cm^3})(8.86 \times 10^{-23} \text{ cm}^3) = 4.81 \times 10^{-22} \text{ g}$$

$$\text{mass of one atom} = \left(\frac{144.24 \text{ g}}{1 \text{ mol}}\right)\left(\frac{1 \text{ mol}}{6.022 \times 10^{23} \text{ atoms}}\right) = 2.40 \times 10^{-22} \text{ g / atom}$$

$$\frac{4.81 \times 10^{-22} \text{ g}}{2.40 \times 10^{-22} \frac{g}{atom}} = 2.00 \text{ atoms}$$

Two atoms are possible only in body-centered cubic unit cells.

9. Under comparable conditions, how much faster will a sample of He effuse through a small opening than a sample of Cl_2?

$$\frac{R_{He}}{R_{Cl_2}} = \sqrt{\frac{M_{Cl_2}}{M_{He}}} = \sqrt{\frac{70.90 \text{ g / mol}}{4.00 \text{ g / mol}}} = 4.21$$

He effuses 4.21 times faster than Cl_2.

Module 13
States of Matter

Introduction

This module describes some laws governing the three states of matter: gas, liquid, and solid. The goals of the module are to:

1. become familiar with how to use the combined and ideal gas laws
2. utilize Graham's law of effusion
3. learn how to determine the relative freezing and boiling points of various liquids based on their intermolecular forces
4. learn to determine the relative melting points of various solids based on the strength of their bonding
5. use the number of particles in the three possible cubic unit cells to calculate atomic radii.

Module 13 Key Equations & Concepts

1. **The combined gas law**

$$\frac{P_1 V_1}{T_1} = \frac{P_2 V_2}{T_2}$$

This is a combination of Boyle's and Charles's gas laws. It is used to determine a new temperature, volume, or pressure of a gas given the original temperature, volume and pressure.

2. **The ideal gas law**

$$PV = nRT$$

This equation is used to calculate the pressure, volume, temperature, or number of moles of a gas given three of the other quantities. It is often used in reaction stoichiometry problems involving gases.

3. **Graham's law of effusion**

$$\frac{R_1}{R_2} = \sqrt{\frac{M_2}{M_1}}$$

This law is used to determine how quickly one gas effuses (or diffuses) relative to another gas. It can also be used to determine the molar masses of gases based on their effusion rates.

4. **Ion-ion interactions, dipole-dipole interactions, hydrogen bonding, London dispersion forces**

These four basic intermolecular forces are important in liquids. The interaction strengths determine liquid boiling points.

5. **Simple Cubic Unit Cells contain 1 particle per unit cell**

The simplest type of cubic unit cell has an atom, ion, or molecule at each of the corners. Because the atoms, ions, and molecules are shared from unit cell to unit cell, each one contributes one-eighth of its volume to a single unit cell. Thus there are 8 x 1/8 =1 atom, ion, or molecule per unit cell.

6. **Body-centered Cubic Unit Cells contain 2 particles per unit cell**

Body-centered cubic unit cells have one more atom, ion, or molecule in the center of the unit cell than in simple cubic unit cells. Thus there are $(8 \times (1/8)) + 1 = 2$ atoms, ions, or molecules per unit cell.

7. **Face-centered Cubic Unit Cells contain 4 particles per unit cell**

Face-centered cubic unit cells have six additional atoms, ions, and molecules (one in each face of the cube) than in a simple cubic unit cell. These atoms, ions, and molecules are shared one-half in each unit cell. Thus there are $(8 \times (1/8)) + (6 \times (1/2)) = 4$ atoms, ions, or molecules per unit cell.

8. **Covalent Network Solids, Ionic solids, Metallic solids, Molecular solids**

There are the four basic types of solids. Bond strength in solids determines their freezing and boiling points.

Sample Exercises
Gas Laws

1. *A sample of a gas initially having a pressure of 1.25 atm and volume of 3.50 L has its volume changed to 7.50 x 10⁴ mL at constant temperature. What is the gas sample's new pressure of the?*

   The correct answer is: 0.0583 atm.

$$7.50 \times 10^4 \text{ mL} \left(\frac{1 \text{ L}}{1000 \text{ mL}} \right) = 75.0 \text{ L}$$

$$\frac{P_1 V_1}{T_1} = \frac{P_2 V_2}{T_2} \text{ simplifies to } P_1 V_1 = P_2 V_2 \text{ at constant temperature } (T_1 = T_2)$$

$$1.25 \text{ atm} \times 3.50 \text{ L} = P_2 \times 75.0 \text{ L}$$

$$\frac{1.25 \text{ atm} \times 3.50 \text{ L}}{75.0 \text{ L}} = P_2$$

$$0.0583 \text{ atm} = P_2$$

 A common mistake students make in gas law problems is to not pay attention to units. Comparable units must be used in these problems. In this problem the two volumes must be in the same units. We must use either mL or L for both V_1 and V_2 but not mixed units..

2. *A gas sample initially having a pressure of 1.75 atm and a volume of 4.50 L at 25.0°C is heated to 37.0°C at a pressure of 1.50 atm. What is the gas's new volume?*
   The correct answer is: 5.46 L.

$$\frac{P_1 V_1}{T_1} = \frac{P_2 V_2}{T_2} \text{ where :}$$

$$P_1 = 1.75\,\text{atm},\ V_1 = 4.50\,\text{L},\ T_1 = 25.0^\circ C = 298.1\,\text{K}$$

$$P_2 = 1.50\,\text{atm and } T_2 = 37.0^\circ C = 310.1\,\text{K}$$

$$V_2 = \frac{P_1 V_1 T_2}{T_1 P_2} = \frac{(1.75\,\text{atm})(4.50\,\text{L})(310.1\,\text{K})}{(298.1\,\text{K})(1.50\,\text{atm})} = 5.46\,\text{L}$$

Gas law problem temperatures must be in Kelvin. Convert temperatures into Kelvin when working any gas law problems.

3. *A gas sample at a pressure of 3.50 atm and temperature of 45.0°C has a volume of 1.65 x 10³ mL. How many moles of gas are in this sample?*
   The correct answer is: 0.221 moles.

$$PV = nRT \text{ where :}$$

$$P = 3.50\,\text{atm},\ V = 1.65 \times 10^3\,\text{mL} = 1.65\,\text{L},\ R = 0.0821\,\frac{\text{L atm}}{\text{mol K}},\ T = 45.0^\circ C = 318.1\,\text{K}$$

$$n = \frac{PV}{RT} = \frac{(3.50\,\text{atm})(1.65\,\text{L})}{\left(0.0821\,\frac{\text{L atm}}{\text{mol K}}\right)(318.1\,\text{K})} = 0.221\,\text{mol}$$

INSIGHT: R is the ideal gas constant. For gas laws, R = 0.0821 L atm/mol K. R defines the units that must be used in ideal gas law problems. P must be in atm, V in L, n in moles, and T in K.

4. *How many grams of CO₂(g) are present in an 11.2 L sample at STP?*
   The correct answer is: 22.0 g.

$$PV = nRT \text{ thus } n = \frac{PV}{RT}$$

$$n = \frac{(1.00\,\text{atm})(11.2\,\text{L})}{\left(0.0821\,\frac{\text{L atm}}{\text{mol K}}\right)(273.15\,\text{K})} = 0.500\,\text{mol}$$

$$0.500\,\text{mol}\left(\frac{44.0\,\text{g CO}_2}{1\,\text{mol CO}_2}\right) = 22.0\,\text{g CO}_2$$

INSIGHT: STP is a symbol for standard temperature and pressure. When you see STP in a gas law problem, the temperature is 273.15 K and the pressure is 1.00 atm or 760 mm Hg.

5. **If 35.0 g of Al react with excess sulfuric acid, how many L of hydrogen gas,**
   H_2, are formed at 1.25 atm and 75.0°C?

$$2\,Al(s)\ +\ 3\,H_2SO_4(aq)\ \rightarrow\ Al_2(SO_4)_3(aq)\ +\ 3\,H_2(g)$$

The correct answer is: 44.6 L.

   a) Calculate the number of moles of hydrogen gas formed in the reaction.

$$(35.0\ \text{g Al})\left(\frac{1\ \text{mol}}{26.98\ \text{g Al}}\right)\left(\frac{3\ \text{mol }H_2}{2\ \text{mol Al}}\right) = 1.95\ \text{mol }H_2$$

   b) Use the ideal gas law to determine the gas volume.

$$75.0°C\ =\ 273.15 + 75.0°C\ =\ 348.1\ K.$$

$$PV\ =\ nRT \text{ thus } V\ =\ \frac{nRT}{P}$$

$$V\ =\ \frac{(1.95\ \text{mol})\left(0.0821\,{}^{L\cdot atm}\!/_{mol\ \cdot K}\right)(348.1\ K)}{1.25\ \text{atm}} = 44.6\ L$$

INSIGHT:	Sample Exercise 5 combines reaction stoichiometry and the ideal gas law to determine the gas volume formed in a reaction.

6. **A gas having a molar mass of 16.0 g/mol effuses through a pinhole 4.00 times**
   faster than an unknown gas. What is the molar mass of the unknown gas?
   The correct answer is: 256 g/mol.

Graham's law of effusion relates the rate at which molecules effuse to the molar masses
of the substances. In this case R_1 = effusion rate of gas 1, R_2 = effusion rate of gas 2,

$$M_1 = \text{molar mass of gas 1, and } M_2 = \text{molar mass of gas 2.}$$

In this problem $R_1 = 4.00$, and $R_2 = 1.00$

$$M_1 = 16.0, \text{ and } M_2 \text{ is unknown.}$$

$$\frac{R_1}{R_2} = \sqrt{\frac{M_2}{M_1}} \text{ thus } \frac{4.00}{1.00} = \sqrt{\frac{M_2}{16.0\ \text{g/mol}}} \text{ to find } M_2 \text{ square both sides of this equation}$$

$$16.0 = \frac{M_2}{16.0\ \text{g/mol}} \text{ thus } M_2 = 16.0 \times 16.0\ \text{g/mol} = 256\ \text{g/mol}$$

INSIGHT:	Deciding which gas has the faster rate is confusing. If you associate the rate with one of the gases, i.e. R_1 with M_1 and R_2 with M_2, the relationship works correctly.

Liquids

7. *Arrange these substances by increasing boiling point.*
   CO_2, $NaCl$, C_2H_5OH, CH_3Cl
   The correct answer is: $CO_2 < CH_3Cl < C_2H_5OH < NaCl$

Boiling points are determined by intermolecular force strengths found in a given liquid. In general, relative intermolecular force strengths are: ion-ion interactions > hydrogen bonding > dipole-dipole interactions > London dispersion forces.

Primary intermolecular forces between molecules of a given substance are determined by compound type and polarity.

Primary Intermolecular Force	Molecule Type
Ion-ion	Ionic compounds
Hydrogen bonding	Molecules with at least one H directly bonded to O, N, or F atom
Dipole-dipole	Polar molecules
London dispersion forces	Nonpolar molecules

Ion-ion interactions are the strongest intermolecular force. Intermolecular force strength decreases going down the above table. Weaker intermolecular forces correlate to a decrease in boiling and/or melting points.

The strongest intermolecular force in liquid CO_2 is London dispersion forces, CH_3Cl's strongest intermolecular forces is dipole-dipole interactions, hydrogen bonding is dominant in C_2H_5OH, and $NaCl$ is an ionic compound. Thus the correct order is:
$CO_2 < CH_3Cl < C_2H_5OH < NaCl$.
Normal boiling points are given below.
CO_2 -78.5 °C
CH_3Cl -24.0 °C
C_2H_5OH 78.0 °C
$NaCl$ 1413 °C

Solids

8. *Lead, Pb, has a density of 11.35 g/cm^3. Solid Pb crystallizes in one of the cubic unit cells with an edge length of 4.95 x 10^{-8} cm. In which of the three unit cells (simple, body-centered, or face-centered) does solid lead, Pb, form crystals? What is the radius, in cm, of a Pb atom?*
   The correct answer is: face-centered cubic with an atomic radius of 1.75 x 10^{-8} cm.

a) First we need to determine the volume of a single unit cell.

For cubes $V = \ell^3$.

$$V = \left(4.95 \times 10^{-8} \text{ cm}\right)^3 = 1.21 \times 10^{-22} \text{ cm}^3$$

b) Use this calculated volume along with the density to determine the mass of a single unit cell.

$$? \text{ g} = 1.21 \times 10^{-22} \text{ cm}^3 \left(\frac{11.35 \text{ g}}{\text{cm}^3}\right) = 1.37 \times 10^{-21} \text{ g}$$

c) Determine the mass of a single Pb atom.

$$? \text{ g} = 207.2 \text{ g/mol} \left(\frac{1 \text{ mol}}{6.022 \times 10^{23} \text{ atoms}}\right) = 3.44 \times 10^{-22} \text{ g/atom}$$

d) Use steps b and c to determine the number of atoms in a single unit cell.

$$? \text{ atoms} = \frac{1.37 \times 10^{-21} \text{ g}}{3.44 \times 10^{-22} \text{ g/atom}} = 3.98 \text{ atoms} \approx 4 \text{ atoms in the unit cell} = \text{ face centered cubic unit cell}$$

Type of unit cell	Particles per unit cell
Simple cubic	1
Body-centered cubic	2
Face-centered cubic	4

e) To calculate the radius of a single Pb atom requires use of the Pythagorean theorem and some knowledge of the geometry of a face - centered cubic unit cell.

(ii)

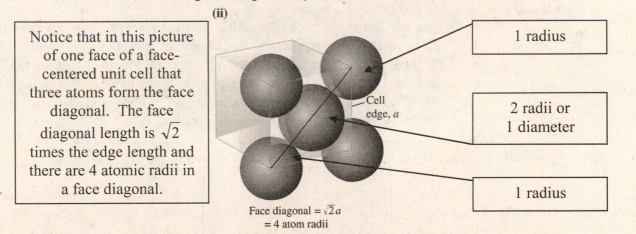

Notice that in this picture of one face of a face-centered unit cell that three atoms form the face diagonal. The face diagonal length is $\sqrt{2}$ times the edge length and there are 4 atomic radii in a face diagonal.

Cell edge, a

1 radius

2 radii or 1 diameter

1 radius

Face diagonal $= \sqrt{2}a$
$= 4$ atom radii

As shown in the diagram : diagonal $= (\sqrt{2})a = 4$ atomi radii

$$\text{diagonal length} = \sqrt{2}\left(4.95 \times 10^{-8} \text{ cm}\right) = 7.00 \times 10^{-8} \text{ cm}$$

$$\text{radius of a Pb atom} = \frac{7.00 \times 10^{-8} \text{ cm}}{4 \text{ radii}} = 1.75 \times 10^{-8} \text{ cm}$$

For other cubic unit cells the geometrical relationships are:
1) **simple cubic unit cells**
   atomic radius = ½ cell edge length
2) **body-centered cubic unit cells**
   atomic radius = $\sqrt{3}$ x ¼ cell edge length

9. *Arrange these substances in order of increasing melting point.*
   CO_2, KCl, Na, SiO_2
   The correct answer is: CO_2 < Na < KCl < SiO_2

INSIGHT:
Melting points of solids are determined by the strength of the force binding substances together. In general, the weakest forces are intermolecular forces found in molecular solids like CO_2, next weakest are metallic bonds as in Na, ionic bonds are relatively strong as found in KCl, and the strongest are covalent bonds from atom to atom bonding network covalent species like SiO_2.

YIELD
The key to melting point problems is determining a solid's classification.
1) **Molecular solids** are covalent compounds having individual molecules. Most of the covalent species you have learned up to now are molecular solids.
2) **Metallic solids** are by far the easiest to classify. Simply look for a metallic element.
3) **Ionic solids** are the basic ionic compounds you have experienced up to this point.
4) The hardest substances to classify are the **network covalent species**. They are covalent species that form extremely large molecules through extended arrays of covalently bonded atoms. Most textbooks have a list of some common network covalent solids that include diamond, graphite, tungsten carbide (WC), and sand (SiO_2). It is probably best to memorize these molecules.

Module 13 relates to some following Modules as shown in the graphic below.

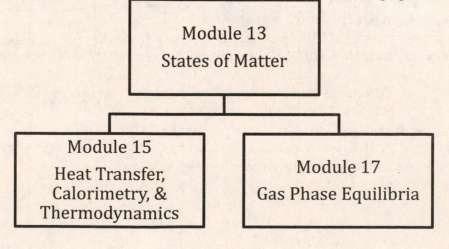

140

Module 14 Predictor Questions

The following questions may help you determine the extent you need to study this module. Questions are ranked according to ability.

        Level 1 = basic proficiency
        Level 2 = mid level proficiency
        Level 3 = high proficiency

If you can correctly answer Level 3 questions you probably do not need to spend much time on this module. If you can only answer Level 1 problems, you should review this module.

Level 1	1. Which of the following compounds are miscible with water? CH_3OH, CH_3COOH, CCl_4, CH_3NH_2, $HOCH_2CH_2OH$
Level 1	2. Which of the following compounds are miscible in hexane? CH_3OH, H_2O, CCl_4, C_8H_{18}, $CaBr_2$
Level 1	3. Choose the statements that are correct, given the following information regarding the solubility of NH_4Cl in water: $$NH_4Cl(s) \rightarrow NH_4Cl(aq) \quad \Delta H_{dissolution} > 0$$

        a) Increasing the water temperature increases the solubility of NH_4Cl in water.
        b) Increasing the water temperature decreases the solubility of NH_4Cl in water.
        c) Increasing the pressure increases the solubility of NH_4Cl in water.
        d) Increasing the pressure decreases the solubility of NH_4Cl in water.
        e) Increasing the pressure will have no effect on the solubility of NH_4Cl in water.

Level 2	4. a) Is CH_3Cl more soluble in CH_3OH or in $CH_3CH_2CH_2CH_2CH_2CH_2OH$? b) Is $HOCH_2CH_2OH$ more soluble in hexane (C_6H_{14}) or water?
Level 2	5. Determine the %w/w and the $X_{sulfuric\ acid}$ of an aqueous H_2SO_4 solution that is 5.00 m. (Water is the solvent.)
Level 2	6. A 16.0% w/w $C_6H_{12}O_6$ solution has a density of 1.0624 g/mL. What is the molar concentration of the solution?

Level 2   7. Sulfur is readily soluble in carbon disulfide, CS_2. The vapor pressure of pure CS_2 is 2.00 atm at 69.1°C. What is the vapor pressure of a solution made by dissolving 32.0 g of S in 380.0 g of CS_2 at 69.1°C?

Level 1   8. Calculate the freezing point, in °C, of a solution containing 8.0 g of sucrose (molar mass = 342 g/mol) in 100. g of water. K_f for water is 1.86 °C/m.

Level 1   9. 60.3 mg of a nonelectrolyte is dissolved in sufficient water at a temperature of 5.0°C to make 150. mL of solution. The osmotic pressure of the solution is 0.200 atm. What is the molar mass of the substance?

Module 14 Predictor Question Solutions

1. Which of the following compounds are miscible with water?
   CH_3OH, CH_3COOH, CCl_4, CH_3NH_2, $HOCH_2CH_2OH$

   Most polar molecules are miscible in water. These polar molecules from the list are miscible in water: CH_3OH, CH_3COOH, CH_3NH_2, $HOCH_2CH_2OH$

2. Which of the following compounds are miscible in hexane?
   CH_3OH, H_2O, CCl_4, C_8H_{18}, $CaBr_2$

   Nonpolar molecules are primarily miscible in nonpolar solvents such as hexane. This includes these nonpolar molecules from the list: CCl_4 and C_8H_{18}

3. Choose the statements that are correct, given the following information regarding the solubility of NH_4Cl in water:

   $$NH_4Cl(s) \rightarrow NH_4Cl(aq) \quad \Delta H_{dissolution} > 0$$

   a) Increasing the water temperature increases the solubility of NH_4Cl in water.
   b) Increasing the water temperature decreases the solubility of NH_4Cl in water.
   c) Increasing the pressure increases the solubility of NH_4Cl in water.
   d) Increasing the pressure decreases the solubility of NH_4Cl in water.
   e) Increasing the pressure will have no effect on the solubility of NH_4Cl in water.

   $\Delta H_{dissolution} > 0$ indicates the reaction is endothermic (heat is a reactant). For an endothermic reaction increasing temperature increases solubility. Changing the pressure has no effect on the solubilities of liquids or solids. Statements a) and e) are true.

4. a) Is CH_3Cl more soluble in CH_3OH or in $CH_3CH_2CH_2CH_2CH_2CH_2OH$
   b) Is $HOCH_2CH_2OH$ more soluble in hexane (C_6H_{14}) or water?

   a) Because CH_3Cl is polar, its solubility is increased in the more polar solvent. Of the two possible solvents both are polar. However, CH_3OH is more polar than the other alcohol because its nonpolar portion (CH_3-) is smaller than the nonpolar portion of the other molecule ($CH_3CH_2CH_2CH_2CH_2CH_2-$). Consequently, CH_3Cl will dissolve more completely in CH_3OH.

   b) $HOCH_2CH_2OH$ is also polar and consequently more soluble in a polar solvent. H_2O is a quite polar solvent. Hexane is nonpolar. Consequently, more $HOCH_2CH_2OH$ will dissolve in H_2O than in hexane.

5. Determine the %w/w and the $X_{sulfuric\ acid}$ of an aqueous H_2SO_4 solution that is 5.00 m. (Water is the solvent.)

Molality, m, is defined as moles of solute per kg of solvent. Therefore, 5.00 m is rewritten as follows:

$$m = \frac{5.00 \text{ mol H}_2\text{SO}_4}{1.00 \text{ kg H}_2\text{O}}$$

$$(5.00 \text{ mol H}_2\text{SO}_4)\left(\frac{98.09 \text{ g H}_2\text{SO}_4}{1 \text{ mol H}_2\text{SO}_4}\right) = 490. \text{ g H}_2\text{SO}_4$$

Mass of solution = 1000 g H$_2$O + 490. g H$_2$SO$_4$ = 1.49 x 10^3 g solution

$$\% \text{ w / w} = \frac{\text{mass solute}}{\text{mass solution}} = \frac{490.45 \text{ g H}_2\text{SO}_4}{1.49 \text{ x } 10^3 \text{ g solution}} = 32.9\% \text{ H}_2\text{SO}_4$$

$$(1000 \text{ g H}_2\text{O})\left(\frac{1 \text{ mol H}_2\text{O}}{18.02 \text{ g H}_2\text{O}}\right) = 55.5 \text{ mol H}_2\text{O}$$

$$X_{\text{sulfuric acid}} = \frac{\text{mol solute}}{\text{mol solute} + \text{mol solvent}} = \frac{5.00 \text{ mol H}_2\text{SO}_4}{5.00 \text{ mol H}_2\text{SO}_4 + 55.5 \text{ mol H}_2\text{O}} = 0.0826$$

6. A 16.0% w/w $C_6H_{12}O_6$ solution has a density of 1.0624 g/mL. What is the molar concentration of the solution?

$$16.0\% \text{ w / w } C_6H_{12}O_6 = \left(\frac{16.0 \text{ g } C_6H_{12}O_6}{100. \text{ g solution}}\right)\left(\frac{1.0624 \text{ g solution}}{1.00 \text{ mL}}\right)\left(\frac{1 \text{ mL}}{10^{-3} \text{ L}}\right)\left(\frac{1 \text{ mol } C_6H_{12}O_6}{180.18 \text{ g } C_6H_{12}O_6}\right)$$

$$16.0\% \text{ w / w } C_6H_{12}O_6 = 0.943 \text{ M}$$

7. Sulfur is readily soluble in carbon disulfide, CS_2. The vapor pressure of pure CS_2 is 2.00 atm at 69.1°C. What is the vapor pressure of a solution made by dissolving 32.0 g of S in 380.0 g of CS_2 at 69.1°C?

$$P_{\text{solution}} = X_{\text{solvent}}P^0_{\text{solvent}}$$
CS_2 is the solvent; S is the solute

$$(32.0 \text{ g S})\left(\frac{1 \text{ mol S}}{32.07 \text{ g S}}\right) = 0.998 \text{ mol S} \qquad (380.0 \text{ g } CS_2)\left(\frac{1 \text{ mol } CS_2}{76.15 \text{ g } CS_2}\right) = 4.99 \text{ mol } CS_2$$

$$X_S = \frac{\text{mol S}}{\text{mol S} + \text{mol } CS_2} = \frac{0.998 \text{ mol}}{0.998 \text{ mol} + 4.99 \text{ mol}} = 0.167$$

$$X_{CS_2} = 1.00 - X_s = 0.833$$

$$P_{\text{solution}} = X_{\text{solvent}}P^0_{\text{solvent}} = (0.833)(2.00 \text{ atm}) = 1.67 \text{ atm}$$

8. Calculate the freezing point, in °C, of a solution containing 8.0 g of sucrose (molar mass = 342 g/mol) in 100. g of water. K_f for water is 1.86 °C/m.

Sucrose is a nonelectrolyte, so $i = 1$

$$\left(8.0 \text{ g sucrose}\right)\left(\frac{1 \text{ mol sucrose}}{342 \text{ g sucrose}}\right) = 2.3 \times 10^{-2} \text{ mol sucrose}$$

100. g H_2O = 0.100 kg H_2O

$$m = \frac{2.3 \times 10^{-2} \text{ mol sucrose}}{0.100 \text{ kg } H_2O} = 0.23 \text{ } m$$

$$\Delta T = iK_f m = \left(1\right)\left(1.86 \text{ °C}/m\right)\left(0.23 \text{ } m\right) = 0.44 \text{ °C}$$

Freezing point of pure H_2O = 0.00 °C

0.00 °C - 0.44 °C = - 0.44 °C

9. 60.3 mg of a nonelectrolyte is dissolved in sufficient water at a temperature of 5.0 °C to make 150. mL of solution. The osmotic pressure of the solution is 0.200 atm. What is the molar mass of the substance?

$$\Pi = MRT$$

$$T = 5.0 \text{ °C} + 273.15 = 278.15 \text{ K}$$

$$M = \frac{\Pi}{RT} = \frac{0.200 \text{ atm}}{\left(0.0821 \frac{L \times atm}{mol \times K}\right)\left(278.1 K\right)} = 0.00876 \text{ } M$$

$$\left(\frac{0.00876 \text{ mol}}{L}\right)\left(0.150 \text{ L}\right) = 0.00131 \text{ mol}$$

$$\text{Molar mass} = \frac{0.0603 \text{ g}}{0.00131 \text{ mol}} = 46.0 \text{ g}/\text{mol}$$

Module 14
Solutions

Introduction
This module discusses solution properties. The primary goals are to:
1. use molecular polarity to predict solubility in various solvents
2. change the solubility of a given species in a solvent
3. convert from one concentration unit to another
4. use Raoult's law to predict the vapor pressure of a solution
5. determine solution freezing and boiling points
6. calculate the osmotic pressure of a solution.

Module 14 Key Equations & Concepts
Like Dissolves Like
This rule states the common phenomenon that polar molecules are readily soluble in other polar molecules and that nonpolar molecules are readily soluble in other nonpolar molecule. However, polar molecules are fairly insoluble in nonpolar molecules.

Solute solubility is increased when the:
1. **solvent is *heated* in an *endothermic* dissolution**
2. **solvent is *cooled* in an *exothermic* dissolution**
3. **pressure of a gas dissolved in a liquid is increased**

Concentration Units
1. **Molarity**

$$M = \frac{\text{moles of solute}}{\text{L of solution}}$$

Molarity is used in reaction stoichiometry and osmotic pressure problems.

2. **Molality**

$$m = \frac{\text{moles of solute}}{\text{kg of solvent}}$$

Molality is used in freezing point depression and boiling point elevation problems.

3. **Percent weight by weight**

$$\% \text{ w/w} = \frac{\text{mass of one solution component}}{\text{mass of total solution}} (100)$$

Percent weight by weight is used for concentrated solutions.

4. **Mole fraction**

$$X_A = \frac{\text{moles of component A}}{\text{total moles of solution}}$$

Mole fraction is ssed in Raoult's Law

5. **Raoult's Law**

$$P_{solution} = X_{solvent} P^0_{solvent}$$

Raoult's law calculates the vapor pressure of a solution containing a nonvolatile solute.

6. **Freezing point depression and Boiling point elevation**

$$\Delta T_f = iK_f m \text{ and } \Delta T_b = iK_b m$$

These relationships describe how much the freezing or boiling temperatures of a solution differ from the pure solvent's freezing and boiling points.

7. **Solution osmotic pressure**

$$\Pi = MRT$$

This equation describes how osmotic pressure depends upon molarity and temperature.

Sample Exercises
Solubility of a Solute in a Given Solvent

1. Which of the following substances are soluble in water?

$SiCl_4$, NH_3, C_8H_{18}, $CaCl_2$, CH_3OH, $Ca_3(PO_4)_2$

The correct answer is: only NH_3, $CaCl_2$, and CH_3OH are soluble in water

The "Like Dissolves Like" rule implies that polar species dissolve in polar species and nonpolar species dissolve in nonpolar species. Consequently, nonpolar species do not dissolve in polar species and polar species do not dissolve in nonpolar species. In this example, NH_3 and CH_3OH are both polar covalent compounds so they dissolve in the highly polar solvent water. $CaCl_2$ is an ionic compound that is water soluble (the solubility rules also apply in these problems). $SiCl_4$ and C_8H_{18} are both nonpolar covalent compounds thus they are insoluble in water. $Ca_3(PO_4)_2$ is an ionic compound that is insoluble in water. Refresh your memory of the solubility rules if necessary.

Keep in mind these two important questions. 1) Are the covalent compounds in the problem polar or nonpolar? 2) Are the ionic compounds in the problem soluble or insoluble based on the solubility rules? Strong acids and bases are also water soluble.

Increasing the Solubility of a Solute in a Given Solvent

2. Given the equation below, which of the following are correct statements?

$$KI(s) \xrightarrow{H_2O} K^+(aq) + I^-(aq) \qquad \Delta H_{dissolution} > 0$$

The correct answer is: only statements a) and f) are true.

a) *Increasing the temperature of the solvent increases the solubility of the compound in the solvent.*

b) *Decreasing the temperature of the solvent increases the solubility of the compound in the solvent.*

147

c) *Changing the temperature of the solvent does <u>not affect</u> the solubility of the compound in the solvent.*

d) <u>*Increasing*</u> *the pressure of the solute <u>increases</u> the solubility of the compound in the solvent.*

e) <u>*Increasing*</u> *the pressure of the solute <u>decreases</u> the solubility of the compound in the solvent.*

f) <u>*Increasing*</u> *the pressure of the solute does <u>not affect</u> the solubility of the compound in the solvent.*

Here are the important hints in this problem that will help you answer it. The KI(s) indicates that this is a solid dissolving in water. Changing the pressure of liquids and solids has no effect on their solubilities. The positive $\Delta H_{dissolution}$ indicates this is an <u>endothermic</u> process. Heating the solvent for endothermic dissolutions increases the solubility of the solute.

$\Delta H_{dissolution} < 0$ is exothermic. $\Delta H_{dissolution} > 0$ is endothermic.

3. *Given the following dissolution in water equation, which of these condition changes are correct statements?*

$$O_2(g) \xrightarrow{\ H_2O\ } O_2(aq) \ \Delta H_{dissolution} < 0$$

The correct answer is: only conditions b) and d) are correct

a) <u>*Increasing*</u> *the temperature of the solvent <u>increases</u> the solubility of the compound in the solvent.*

b) <u>*Decreasing*</u> *the temperature of the solvent <u>increases</u> the solubility of the compound in the solvent.*

c) *Changing the temperature of the solvent does <u>not affect</u> the solubility of the compound in the solvent.*

d) <u>*Increasing*</u> *the pressure of the solute <u>increases</u> the solubility of the compound in the solvent.*

e) <u>*Increasing*</u> *the pressure of the solute <u>decreases</u> the solubility of the compound in the solvent.*

f) <u>*Increasing*</u> *the pressure of the solute does <u>not affect</u> the solubility of the compound in the solvent.*

Important hints in this problem are 1) $O_2(g)$ indicates this is a gas dissolving in water. Increasing gas pressure has a significant effect on gas solubility. In general, increasing the pressure of a gas increases its solubility in a liquid. 2) The negative $\Delta H_{dissolution}$ indicates that this is an <u>exothermic</u> process. Heating the solvent for exothermic dissolutions decreases the solubility of the solute. Cooling the solvent increases the solubility of the solute in exothermic dissolutions.

Concentration Unit Conversion

4. An aqueous sulfuric acid solution is 3.75 M with a density of 1.225 g/mL. What is the concentration of this solution in molality (m), percent weight by weight (% w/w) of H_2SO_4, and mole fraction ($X_{sulfuric\ acid}$) of H_2SO_4?

The correct answers are: 4.38 m, 30.0 % w/w, and $X_{sulfuric\ acid}$ = 0.0730

$$3.75\ M\ H_2SO_4 = \frac{3.75\ \text{moles of}\ H_2SO_4}{1.00\ \text{L of solution}}$$

Solute and solvent masses must be separated to determine the other concentrations. Molarity tells us the solution volume not the mass. The solution density helps us calculate its mass. To make the calculation as easy as possible, we can assume we have one liter of 3.75 M H_2SO_4 solution.

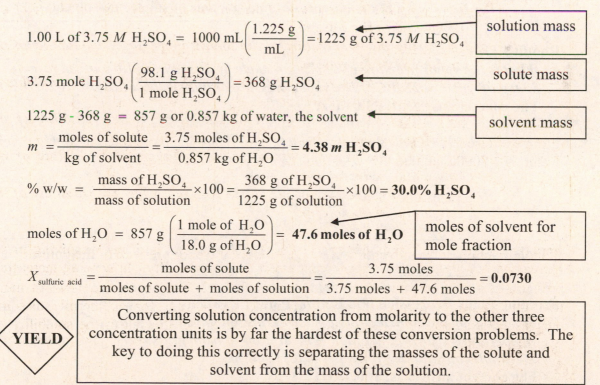

$$1.00\ \text{L of}\ 3.75\ M\ H_2SO_4 = 1000\ \text{mL}\left(\frac{1.225\ \text{g}}{\text{mL}}\right) = 1225\ \text{g of}\ 3.75\ M\ H_2SO_4$$ → solution mass

$$3.75\ \text{mole}\ H_2SO_4\left(\frac{98.1\ \text{g}\ H_2SO_4}{1\ \text{mole}\ H_2SO_4}\right) = 368\ \text{g}\ H_2SO_4$$ → solute mass

$$1225\ \text{g} - 368\ \text{g} = 857\ \text{g or}\ 0.857\ \text{kg of water, the solvent}$$ → solvent mass

$$m = \frac{\text{moles of solute}}{\text{kg of solvent}} = \frac{3.75\ \text{moles of}\ H_2SO_4}{0.857\ \text{kg of}\ H_2O} = \mathbf{4.38\ m\ H_2SO_4}$$

$$\%\ \text{w/w} = \frac{\text{mass of}\ H_2SO_4}{\text{mass of solution}} \times 100 = \frac{368\ \text{g of}\ H_2SO_4}{1225\ \text{g of solution}} \times 100 = \mathbf{30.0\%\ H_2SO_4}$$

$$\text{moles of}\ H_2O = 857\ \text{g}\left(\frac{1\ \text{mole of}\ H_2O}{18.0\ \text{g of}\ H_2O}\right) = \mathbf{47.6\ moles\ of\ H_2O}$$ → moles of solvent for mole fraction

$$X_{sulfuric\ acid} = \frac{\text{moles of solute}}{\text{moles of solute + moles of solution}} = \frac{3.75\ \text{moles}}{3.75\ \text{moles} + 47.6\ \text{moles}} = \mathbf{0.0730}$$

⟨YIELD⟩ Converting solution concentration from molarity to the other three concentration units is by far the hardest of these conversion problems. The key to doing this correctly is separating the masses of the solute and solvent from the mass of the solution.

5. An aqueous sucrose, $C_{12}H_{22}O_{11}$, solution is 11.0 % w/w with a density of 1.0432 g/mL. What is the concentration of this solution in molarity (M), molality (m), and mole fraction ($X_{sucrose}$) of $C_{12}H_{22}O_{11}$?

The correct answer is 0.335 *M*, 0.361 *m*, and 0.00646 $X_{sucrose}$

Assuming we have 100.0 g of solution, we know there is 11.0 g of sucrose and 89.0 g or 0.0890 kg of water. This is one key to solving this problem.

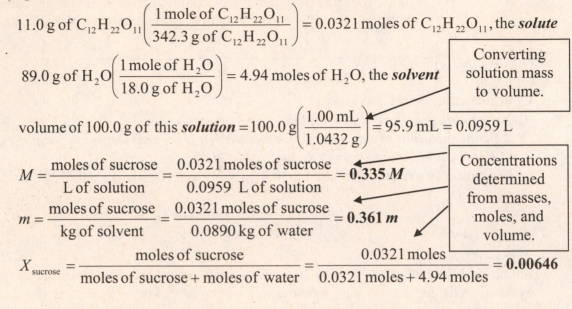

$$11.0 \text{ g of } C_{12}H_{22}O_{11}\left(\frac{1 \text{ mole of } C_{12}H_{22}O_{11}}{342.3 \text{ g of } C_{12}H_{22}O_{11}}\right) = 0.0321 \text{ moles of } C_{12}H_{22}O_{11}, \text{ the } \textbf{\textit{solute}}$$

$$89.0 \text{ g of } H_2O\left(\frac{1 \text{ mole of } H_2O}{18.0 \text{ g of } H_2O}\right) = 4.94 \text{ moles of } H_2O, \text{ the } \textbf{\textit{solvent}}$$

Converting solution mass to volume.

$$\text{volume of 100.0 g of this } \textbf{\textit{solution}} = 100.0 \text{ g}\left(\frac{1.00 \text{ mL}}{1.0432 \text{ g}}\right) = 95.9 \text{ mL} = 0.0959 \text{ L}$$

Concentrations determined from masses, moles, and volume.

$$M = \frac{\text{moles of sucrose}}{\text{L of solution}} = \frac{0.0321 \text{ moles of sucrose}}{0.0959 \text{ L of solution}} = \textbf{0.335 } \textit{M}$$

$$m = \frac{\text{moles of sucrose}}{\text{kg of solvent}} = \frac{0.0321 \text{ moles of sucrose}}{0.0890 \text{ kg of water}} = \textbf{0.361 } \textit{m}$$

$$X_{sucrose} = \frac{\text{moles of sucrose}}{\text{moles of sucrose} + \text{moles of water}} = \frac{0.0321 \text{ moles}}{0.0321 \text{ moles} + 4.94 \text{ moles}} = \textbf{0.00646}$$

INSIGHT: Exercise 5 is easier than exercise 4 because % w/w is a concentration unit that easily separates into amounts of solute and solvent.

Raoult's Law
6. *What is the vapor pressure of a solution made by dissolving 11.0 g of sucrose in 89.0 g of water at 25.0°C? The vapor pressure of pure water at 25.0°C is 23.76 torr.*
   The correct answer is: 23.60 torr

From exercise 5, we know that $X_{sucrose} = 0.00646$. Thus, the mole fraction of the solvent, water, is $1.00000 - 0.00646 = 0.99354$

$$P_{solution} = X_{solvent} P^0_{solvent}$$
$$P_{solution} = 0.99354 \, (23.76 \text{ torr}) = 23.60 \text{ torr}$$

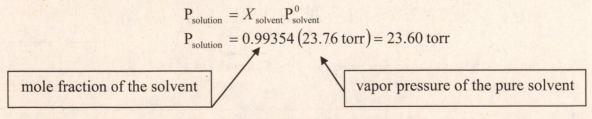

mole fraction of the solvent   vapor pressure of the pure solvent

Freezing Point Depression and Boiling Point Elevation
7. *If 11.0 g of the nonelectrolyte sucrose are dissolved in 89.0 g of water, at what temperature will this solution boil at 1.00 atm of pressure? The boiling point elevation constant, K_b, for water is 0.512 °C/m.*
   The correct answer is: 100.185 °C.

From exercise 5 we know this solution concentration is 0.361 *m*.

The **van't Hoff factor, *i*,** indicates the extent to which the solute dissociates. For an ionic compound, *i*, is close to the number of ions in the compound. For nonelectrolytes, *i* is 1.

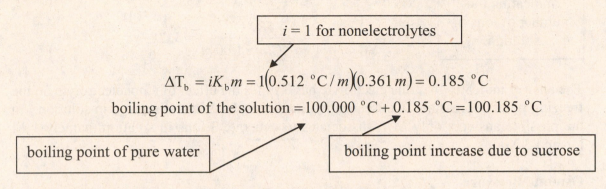

$$i = 1 \text{ for nonelectrolytes}$$

$$\Delta T_b = iK_b m = 1(0.512 \ ^\circ C/m)(0.361 \ m) = 0.185 \ ^\circ C$$

$$\text{boiling point of the solution} = 100.000 \ ^\circ C + 0.185 \ ^\circ C = 100.185 \ ^\circ C$$

boiling point of pure water

boiling point increase due to sucrose

8. **12.4 g of a nonelectrolyte are dissolved in 100.0 g of water then the solution is frozen. The freezing point of the solution is -5.00 °C. What is the molar mass of the nonelectrolyte? The freezing point depression constant, K_f, for water is 1.86 °C/m.**

The correct answer is: 46.1 g/mol.

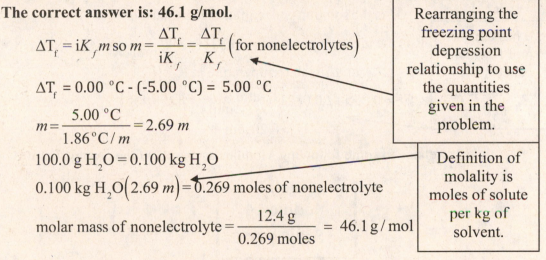

$$\Delta T_f = iK_f m \text{ so } m = \frac{\Delta T_f}{iK_f} = \frac{\Delta T_f}{K_f} \left(\text{for nonelectrolytes} \right)$$

$$\Delta T_f = 0.00 \ ^\circ C - (-5.00 \ ^\circ C) = 5.00 \ ^\circ C$$

$$m = \frac{5.00 \ ^\circ C}{1.86 \ ^\circ C/m} = 2.69 \ m$$

$$100.0 \text{ g } H_2O = 0.100 \text{ kg } H_2O$$

$$0.100 \text{ kg } H_2O(2.69 \ m) = 0.269 \text{ moles of nonelectrolyte}$$

$$\text{molar mass of nonelectrolyte} = \frac{12.4 \text{ g}}{0.269 \text{ moles}} = 46.1 \text{ g/mol}$$

Rearranging the freezing point depression relationship to use the quantities given in the problem.

Definition of molality is moles of solute per kg of solvent.

9. **A 1.00 m solution of a strong electrolyte dissolved in 100.0 g of water has a freezing point of -5.58°C. Which of these generic ionic formulas corresponds to the electrolyte formula? (M represents a typical metal cation and X a typical anion. The freezing point depression constant, K_f, for water is 1.86 °C/m.)**

The correct answer is: b) MX₂

a) *MX*
b) *MX₂*
c) *MX₃*
d) *M₂X₃*
e) *M₂X₄*

Electrolytes have an i value > 1 indicating the number of ions in solution.

$\Delta T_f = iK_f m$ which can be rearranged to $\dfrac{\Delta T_f}{K_f m} = i$

$\Delta T_f = 0.00\ ^\circ C - (-5.58\ ^\circ C) = 5.58\ ^\circ C$

$i = \dfrac{\Delta T_f}{K_f m} = \dfrac{5.58\ ^\circ C}{1.86\ ^\circ C/m \times 1.00\ m} = 3.00$

The answer indicates that this electrolyte has 3 times the effect of a nonelectrolyte on the freezing point depression. Consequently, there must be 3 ions dissolved in solution. In the possible answers only MX$_2$ dissociates to generate 3 ions in solution, namely 1 M ion and 2 X ions.

Osmotic Pressure

10. What is the osmotic pressure of a 0.335 M sucrose solution at 25.0°C?
   The correct answer is: 8.20 atm.

$\Pi = M$RT where Π is the osmotic pressure in atm,

M is the solution concentration in molarity,

M has units of mol/L which cancels with the L/mol units in R.

R is the ideal gas constant $= 0.0821 \dfrac{L \cdot atm}{mol \cdot K}$

and T is the temperature in K.

$\Pi = 0.335\ M\left(0.0821 \dfrac{L \cdot atm}{mol \cdot K}\right)(298.1\,K)$

$\Pi = 8.20\ atm$

Module 14 relates to some following Modules as shown in the graphic below.

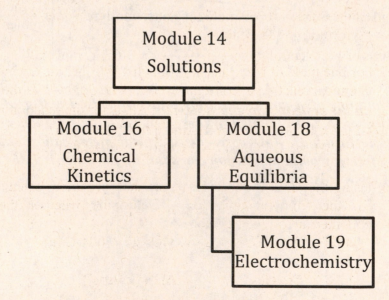

Module 15 Predictor Questions

The following questions may help you determine the extent you need to study this module. Questions are ranked according to ability.

Level 1 = basic proficiency
Level 2 = mid level proficiency
Level 3 = high proficiency

If you can correctly answer Level 3 questions you probably do not need to spend much time on this module. If you can only answer Level 1 problems, you should review this module.

Level 1   1. A chemical reaction releases 58,500 J of heat into 150 g of water. Assuming no heat is lost to the surroundings, what is the water temperature *increase*? The specific heat of liquid water is 4.184 J/g·°C.

Level 1   2. A 25.0 g sample of Indium (In) is heated by exposure to 1.50×10^3 J. The In temperature rises by 258 °C. What is the specific heat of In in J/g·°C?

Level 1   3. Calculate the amount of heat required to heat 10.0 grams of ice at -20.0°C to 120.°C. The specific heats are: $H_2O(s)$ = 2.09 J/g·°C; $H_2O(l)$ = 4.18 J/g·°C; $H_2O(g)$ = 2.03 J/g·°C. The heats of fusion and vaporization are, respectively: $H_2O(s)$ = 333 J/g; $H_2O(l)$ = 2260 J/g.

Level 1   4. What is the change in internal energy of a system, in J, if the system emits 763 J of work to its surroundings while 655 J of work is performed on the system?

Level 2   5. For one mole of reactions, how much work (in J) is done by this chemical reaction at constant pressure and 32.0°C? Is the work done *on the system* or *by the system?* If ΔH_{rxn} for the reaction is -2219.8 kJ (-2.22×10^6 J), what are ΔE and q for this system?
$$C_3H_8(g) + 5\ O_2(g) \rightarrow 3\ CO_2(g) + 4\ H_2O(g)$$

Level 1   6. Using the table of thermodynamic data provided, calculate the ΔH^0_{rxn} for the following reaction. Determine whether the reaction is endothermic or exothermic.
$$2\ C_6H_6\ (l) + 15\ O_2(g) \rightarrow 12\ CO_2(g) + 6\ H_2O\ (g)$$

Species	ΔH^0_f (kJ/mol)
C_6H_6	49.04
CO_2	-393.5
H_2O	-241.8

Level 1   7. Determine ΔS^0_{rxn}, in J/mol, for the combustion of 1 mol of $C_3H_{8\,(g)}$ at 25°C.

$$C_3H_8(g) + 5\, O_2(g) \rightarrow 3\, CO_2(g) + 4\, H_2O(g)$$

Species	ΔS^0_f (J/mol·K)
C_3H_8	269.9
CO_2	213.7
H_2O	69.9
O_2	205.0

Level 1   8. For a certain process at 127°C, ΔG = -16.20 kJ and ΔH = -17.0 kJ. What is the entropy change, in J/K, for this process at 127°C?

Level 1   9. Using the data given below, determine the ΔG^0_{rxn} for this chemical reaction.   Determine whether the reaction is spontaneous or nonspontaneous.

$$CS_2(g) + 3\, O_2(g) \rightarrow CO_2(g) + 2\, SO_2(g)$$

Species	ΔG^0_f (kJ/mol)
CS_2	67.15
O_2	0.0
CO_2	-394.4
SO_2	-300.2

Module 15 Predictor Question Solutions

1. A chemical reaction releases 58,500 J of heat into 150 g of water. Assuming no heat is lost to the surroundings, what is the water temperature *increase*? The specific heat of liquid water is 4.184 J/g·°C.

$$q = mC\Delta T \Rightarrow \Delta T = \frac{q}{mC} = \frac{58500 \text{ J}}{(150.\,\text{g})(4.184\frac{\text{J}}{\text{g}\times^\circ \text{C}})} = 93.2^\circ\text{C}$$

2. A 25.0 g sample of Indium (In) is heated by exposure to 1.50×10^3 J. The In temperature rises by 258 °C. What is the specific heat of In in J/g·°C?

$$q = mC\Delta T \Rightarrow C = \frac{q}{m\Delta T} = \frac{1.50 \times 10^3 \text{ J}}{(25.0\,\text{g})(258^\circ\text{C})} = 0.233\frac{\text{J}}{\text{g}\times^\circ \text{C}}$$

3. Calculate the amount of heat required to heat 10.0 grams of ice at -20.0°C to 120.°C. The specific heats are: $H_2O(s) = 2.09$ J/g·°C; $H_2O(l) = 4.18$ J/g·°C; $H_2O(g) = 2.03$ J/g·°C. The heats of fusion and vaporization are, respectively: $H_2O(s) = 333$ J/g; $H_2O(l) = 2260$ J/g.

Five steps are required: a) heat ice from -20.0°C to 0.00 °C; b) melt ice; c) heat water from 0.00 °C to 100.0 °C; d) evaporate water; e) heat steam from 100.0 °C to 120.0 °C.

a) $q = mC\Delta T = (10.0\,\text{g})(2.09\frac{\text{J}}{\text{g}\times^\circ \text{C}})(20.0^\circ\text{C}) = 418\text{ J}$

b) $q = m\Delta H_{fusion} = (10.0\,\text{g})(333\frac{\text{J}}{\text{g}}) = 3330\text{ J}$

c) $q = mC\Delta T = (10.0\,\text{g})(4.18\frac{\text{J}}{\text{g}\times^\circ \text{C}})(100.^\circ\text{C}) = 4180\text{ J}$

d) $q = m\Delta H_{evaporation} = (10.0\,\text{g})(2260\frac{\text{J}}{\text{g}}) = 22600\text{ J}$

e) $q = mC\Delta T = (10.0\,\text{g})(2.03\frac{\text{J}}{\text{g}\times^\circ \text{C}})(10.0^\circ\text{C}) = 203\text{ J}$

Sum of five steps = 3.07×10^4 J = 30.7 kJ

4. What is the change in internal energy of a system, in J, if the system emits 763 J of work to its surroundings while 655 J of work is performed on the system?

$\Delta E = q + w = \text{-763 J} + (655\text{ J}) = \text{-108 J}$

5. For one mole of reactions, how much work (in J) is done by this chemical reaction at constant pressure and 32.0°C? Is the work done *on the system* or *by the system?* If ΔH_{rxn} for the reaction is -2219.8 kJ (-2.22 x 10^6 J), what are ΔE and q for this system?

$$C_3H_8(g) + 5\ O_2(g) \rightarrow 3\ CO_2(g) + 4\ H_2O(g)$$

$\Delta n = \Sigma$ **mol gas (product) -** Σ **mol gas (reactant) = 7 mol - 6 mol = 1 mol**

$T = 32.0^\circ C + 273.15 = 305.1\ K$

$w = -\Delta n_{gas} RT = -(1\ mol)(8.314 \dfrac{J}{mol \times K})(305.1\ K) = -2.54 \times 10^3\ J$

The negative sign indicates that work is done *by* the system.

$\Delta H_{rxn} = q$ **(at constant pressure) = - 2219.8 kJ)**

$\Delta E = w + q = -2.54 \times 10^3\ J + -2.22 \times 10^6\ J = -2.22 \times 10^6\ J$

6. Using the table of thermodynamic data provided, calculate the ΔH^0_{rxn} for the following reaction. Determine whether the reaction is endothermic or exothermic.

$$2\ C_6H_6\ (l) + 15\ O_2(g) \rightarrow 12\ CO_2(g) + 6\ H_2O\ (g)$$

Species	ΔH^0_f (kJ/mol)
C_6H_6	49.04
CO_2	-393.5
H_2O	-241.8

$\Delta H^0_{rxn} = \Sigma n\ \Delta H^0_{f\ products} - \Sigma n\ \Delta H^0_{f\ reactants}$

$\Delta H^0_{rxn} = [12(-395.5\ kJ\ /\ mol) + 6(-241.8\ kJ\ /\ mol)] - [2(49.04\ kJ\ /\ mol)]$

$\Delta H^0_{rxn} = -6.271 \times 10^3\ kJ$

6. Determine ΔS^0_{rxn}, in J/mol, for the combustion of 1 mol of $C_3H_{8\ (g)}$ at 25°C.

$$C_3H_8(g) + 5\ O_2(g) \rightarrow 3\ CO_2(g) + 4\ H_2O(g)$$

Species	ΔS^0_f (J/mol·K)
C_3H_8	269.9
CO_2	213.7
H_2O	69.9
O_2	205.0

$$\Delta S^0_{rxn} = \Sigma \, n \, S^0_{f \, products} - \Sigma \, n \, S^0_{f \, reactants}$$

$$\Delta S^0_{rxn} = [3(213.7 \text{ J / mol} \times \text{K}) + 4(69.9 \text{ J / mol} \times \text{K})] -$$

$$[5(205.0 \text{ J / mol} \times \text{K}) + 1(269.9 \text{ J / mol} \times \text{K})]$$

$$\Delta S^0_{rxn} = -374.2 \text{ J / K}$$

8. For a certain process at 127°C, $\Delta G = -16.20$ kJ and $\Delta H = -17.0$ kJ. What is the entropy change, in J/K, for this process at 127°C?

$$\Delta G = \Delta H - T\Delta S \Rightarrow \Delta S = \frac{-\Delta G + \Delta H}{T}$$

$$T = 127^\circ C + 273.15 = 400. \text{ K}$$

$$\Delta S = \frac{-(-16.20 \text{ kJ}) + (-17.0 \text{ kJ})}{400. \text{ K}} = -2.00 \times 10^{-3} \text{ kJ / K}$$

9. Using the data given below, determine ΔG^0_{rxn} for this chemical reaction. Determine whether the reaction is spontaneous or nonspontaneous.

$$CS_2(g) + 3 \, O_2(g) \rightarrow CO_2(g) + 2 \, SO_2(g)$$

Species	ΔG^0_f (kJ/mol)
CS_2	67.15
O_2	0.0
CO_2	-394.4
SO_2	-300.2

$$\Delta G^0_{rxn} = \sum n\Delta G^0_{f \, products} - \sum n\Delta G^0_{f \, reactants}$$

$$\Delta G^0_{rxn} = \left[(-394.4) + 2(-300.2) \right] - \left[(67.15) + 3(0.0) \right] \text{ kJ / mol rxn}$$

$$\Delta G^0_{rxn} = -1061.95 \text{ kJ / mol rxn}$$

A negative value for ΔG^0_{rxn} indicates the reaction is spontaneous .

Module 15
Heat Transfer, Calorimetry, and Thermodynamics

Introduction

This module presents a brief discussion of chemically related heat topics. The following major topics must be addressed. All focus on how energy is transferred from one chemical system to another. The primary goals are to understand:

1. basic heat transfer equation and its impact on both calorimetry and heating substances that remain in a single phase
2. simple chemical thermodynamics including the change in energy (ΔE) of a system
3. heat (q) and work (w) involved in an energy change
4. change in enthalpy (ΔH)
5. calculation of ΔH using Hess's law
6. calculation of entropy changes (ΔS) and Gibbs Free Energy (ΔG)
7. temperature dependence of Gibbs Free Energy change

Module 15 Key Equations & Concepts

1. $q = mC\Delta T$

This heat transfer equation allows us to calculate the energy emitted or absorbed when an object is warmed or cooled. (q is the heat, m is the mass, C is the specific heat, and ΔT is the temperature change.) It is used in calorimetry, heat lost = heat gained problems, and to determine the heat necessary to heat or cool a substance remaining *in a single phase*.

2. $\Delta E = q + w$

Energy change for a chemical system is determined by two factors, 1) how much heat (q) enters or leaves the system and 2) how much work (w) the system does in expanding or contracting against a constant pressure, typically atmospheric pressure. Understanding the sign conventions (i.e., +q or –q) for heat and work are crucial in these problems.

3. $w = -P\Delta V = -\Delta n_{gas}RT$ (at constant temperature and pressure)

This relationship defines the work a system does at constant temperature and pressure. (w is the work, P is the pressure, ΔV is the volume change, Δn_{gas} is the change in the number of moles of gas, R is the ideal gas constant, and T is the temperature.) It also describes the work that a system does or has done on it when the number of moles of gas changes.

4. $\Delta H = \Delta E + P\Delta V = q_P$ at constant temperature and pressure

This definition of enthalpy, ΔH, establishes its relationship to the system energy change. (q_P is the heat flow at constant pressure.)

5. $\Delta H^0_{rxn} = \sum n \Delta H^0_{f \; products} - \sum n \Delta H^0_{f \; reactants}$

   Hess's law determines the heat absorbed or released by a chemical reaction from formation enthalpies of products and reactants. ($\Delta H^0_{f \; products}$ is the enthalpy of formation of the product substances at standard conditions. $\Delta H^0_{f \; reactants}$ is the enthalpy of formation of the reactant substances at standard conditions. n represents the stoichiometric coefficients in the balanced chemical reaction.)

6. $\Delta S^0_{rxn} = \sum n S^0_{f \; products} - \sum n S^0_{f \; reactants}$

   This relationship determines the entropy change for a chemical reaction from the standard entropies of formation for the products and reactants at standard conditions.

7. $\Delta G^0_{rxn} = \sum n G^0_{f \; products} - \sum n G^0_{f \; reactants}$

   This relationship determines the Gibbs free energy change for a chemical reaction given standard free energies of formation for the products and reactants under standard conditions.

8. $\Delta G = \Delta H - T \Delta S$

   This is the definition of Gibbs Free Energy. Among other things it is used to determine the Gibbs Free Energy temperature dependence.

Sample Exercises
Heat Transfer Calculations

1. *How much heat is required to heat 75.0 g of aluminum, Al, from 25.0°C to 175.0°C? The specific heat of Al is 0.900 J/g °C.*
   The correct answer is: 1.01 x 10⁴ J or 10.1 kJ.

heat required	mass	specific heat	temperature change

$$q = m \; C \; \Delta T$$
$$= 75.0 \, g (0.900 \, J/g°C)(175.0 - 25.0°C)$$

Final temperature – Initial temperature

$$= 67.5 \, J/°C (150.0°C)$$
$$= 1.01 \times 10^4 \, J \text{ or } 10.1 \, kJ$$

q is positive indicating that Al **absorbs** heat.

CAUTION

When calculating ΔT always use $T_{final} - T_{initial}$ to ensures that q has the correct sign.

2. *A 75.0 g piece of aluminum, Al, initially at 175.0 °C is dropped into a coffee cup calorimeter containing 150.0 g of H₂O initially at 15.0 °C. Calculate the system temperature at thermal equilibrium. Assume no heat is lost to the calorimeter. Specific heats of Al and water are 0.900 J/g °C and 4.18 J/g°C, respectively. The correct answer is: 30.6°C.*

$$\text{heat lost by the Al} = \text{heat gained by the H}_2\text{O}$$

$$-m_{Al}C_{Al}\Delta T_{Al} = m_{H_2O}C_{H_2O}\Delta T_{H_2O}$$

$$-(75.0 \text{ g})(0.900 \text{ J/g}^\circ\text{C})(T_{final} - 175.0^\circ\text{C}) = (150.0 \text{ g})(4.18 \text{ J/g}^\circ\text{C})(T_{final} - 15.0^\circ\text{C})$$

$$(-67.5 \, T_{final} + 11{,}812.5 \,)\, J = (627.0 \, T_{final} - 9{,}405.0)\, J$$

$$11{,}812.5 + 9405.0 = (627.0 \, T_{final} + 67.5 \, T_{final})$$

$$21{,}217.5 = 694.5 \, T_{final}$$

$$\frac{21{,}217.5}{694.5} = T_{final}$$

$$30.6^\circ\text{C} = T_{final}$$

The Joules units cancel.

INSIGHT:	Typical heat lost = heat gained problems involve mixing of two substances at different initial temperatures in a common container. If no heat is lost to the surroundings, then all heat lost by one substance is gained by the other and final temperatures of both substances are equal. Set up the problem so that the heat transfer equations are equal but have opposite signs. You may be asked to determine the specific heat of one of the substances or the final temperature. Example 2 is a form of the latter problem and is the harder of the two types.

3. *How much heat is required to convert 150.0 g of solid Al at 458°C into liquid Al at 758°C? The normal melting point of Al is 658°C. Specific heats for Al are, C_{solid} = 24.3 J/mol °C and C_{liquid} = 29.3 J/mol °C. ΔH_{fusion} for Al = 10.6 kJ/mol. The correct answer is: q = 102.2 kJ.*

INSIGHT:	This problem requires three separate calculations. The final answer is the sum of those three calculations. The steps are: 1) heat required to warm Al from 458°C to its melting point, 658°C, 2) heat required to melt Al, and 3) heat required to heat liquid Al from its melting point to 758°C. These steps are illustrated in the diagram below.

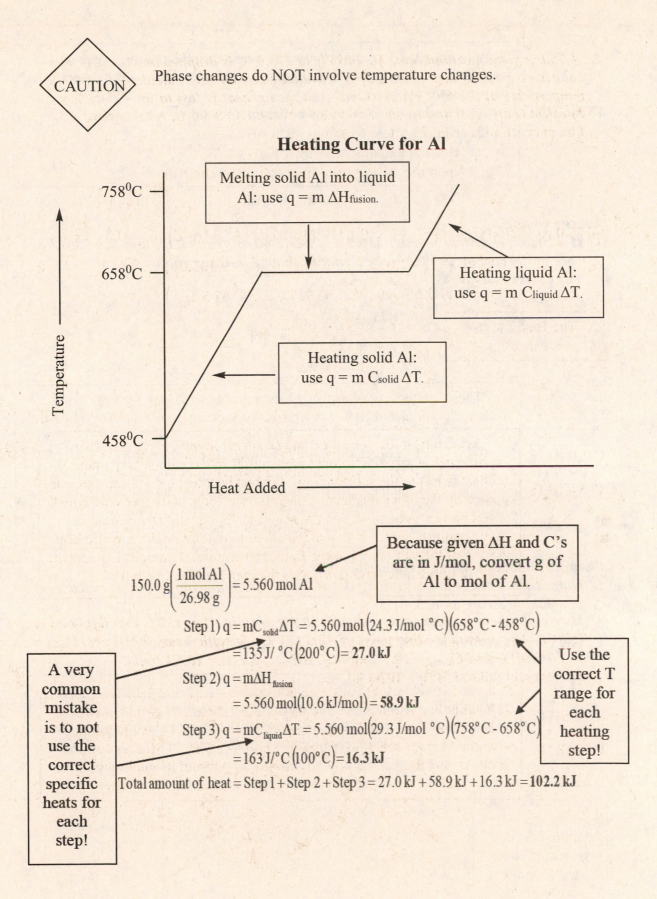

CAUTION   Phase changes do NOT involve temperature changes.

Heating Curve for Al

758⁰C — Melting solid Al into liquid Al: use $q = m\,\Delta H_{fusion}$.

658⁰C — Heating liquid Al: use $q = m\,C_{liquid}\,\Delta T$.

Heating solid Al: use $q = m\,C_{solid}\,\Delta T$.

458⁰C —

Temperature

Heat Added

Because given ΔH and C's are in J/mol, convert g of Al to mol of Al.

$$150.0\,\text{g}\left(\frac{1\,\text{mol Al}}{26.98\,\text{g}}\right) = 5.560\,\text{mol Al}$$

Step 1) $q = mC_{solid}\Delta T = 5.560\,\text{mol}\,(24.3\,\text{J/mol}\,°C)(658°C - 458°C)$

$= 135\,\text{J/}°C\,(200°C) = \textbf{27.0 kJ}$

Step 2) $q = m\Delta H_{fusion}$

$= 5.560\,\text{mol}(10.6\,\text{kJ/mol}) = \textbf{58.9 kJ}$

Step 3) $q = mC_{liquid}\Delta T = 5.560\,\text{mol}(29.3\,\text{J/mol}\,°C)(758°C - 658°C)$

$= 163\,\text{J/}°C\,(100°C) = \textbf{16.3 kJ}$

Total amount of heat = Step 1 + Step 2 + Step 3 = 27.0 kJ + 58.9 kJ + 16.3 kJ = **102.2 kJ**

A very common mistake is to not use the correct specific heats for each step!

Use the correct T range for each heating step!

161

Sample exercise 3 involves only one phase change, melting solid Al. If a second phase change were included, boiling liquid Al into gaseous Al, these steps must be included.
1) Heating liquid Al to the boiling point using $q = m\, C_{liquid}\, \Delta T$.
2) Boiling liquid Al using $q = m\, \Delta H_{vaporization}$.
3) Heating gaseous Al using $q = m\, C_{gas}\, \Delta T$.

Energy Change

4. *If a chemical system releases 350.0 J of heat to its surroundings and has 75.0 J of work performed on it, what is the resulting change in energy of the system?*
   The correct answer is: -275.0 J.

$$\Delta E = q + w$$
$$= -350.0\ J + 75.0\ J$$
$$= -275.0\ J$$

INSIGHT: These are the easiest energy change problems. Look for heat being released (-) or absorbed (+) as well as work being done on (+) or by (-) the system.
__You must know the following sign conventions.__
q > 0 indicates that heat is **absorbed** by the system
q < 0 indicates that heat is **released** by the system
w > 0 indicates that work is done **on** the system
w < 0 indicates that work is done **by** the system

Chemical Reaction Work Relationship

5. *How much work is done on or by the system in the following chemical reaction at constant pressure and 25.0°C?*
   $$C_2H_5OH(l)\ +\ 3\ O_2(g) \rightarrow 2\ CO_2(g)\ +\ 3\ H_2O(g)$$
   The correct answer is: -4.957 x 10³ J; work is being done __by__ the system

INSIGHT: Look for a question with a chemical reaction having a change in the number of moles of gaseous reactants and products that asks for the amount of work performed. Be sure you determine the change in the number of moles of gas as follows:
$$\Delta n_{gas} = \Sigma \text{moles of gas}_{products} - \Sigma \text{moles of gas}_{reactants}$$

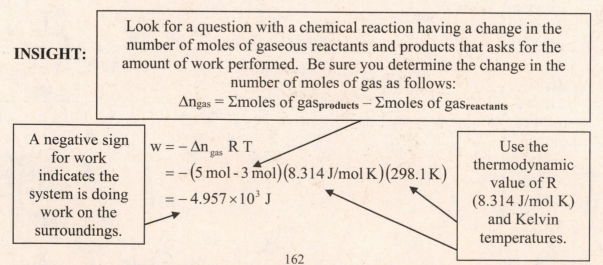

A negative sign for work indicates the system is doing work on the surroundings.

$$w = -\Delta n_{gas}\, R\, T$$
$$= -(5\ mol - 3\ mol)(8.314\ J/mol\ K)(298.1\ K)$$
$$= -4.957 \times 10^3\ J$$

Use the thermodynamic value of R (8.314 J/mol K) and Kelvin temperatures.

Enthalpy Change to Energy Change Relationship

6. *Given the following information about this chemical reaction at constant pressure and temperature of 25.0°C, what are the values of ΔE, q, and w for this reaction?*

$$C_2H_5OH(l) \ + \ 3 \ O_2(g) \rightarrow 2 \ CO_2(g) \ + \ 3 \ H_2O(g) \quad \Delta H^0_{rxn} = -1234.7 \ kJ/mol \ rxn$$

The correct answer is: q = -1234.7 kJ, w = -4.957 kJ, and ΔE = -1239.3 kJ.

Exercise 5 tells us that the value of w for this reaction is -4.957×10^3 J or -4.957 kJ. The definition of enthalpy change, $\Delta H = \Delta E + P\Delta V = q_P$ gives us a method to determine q.

The ΔH^0_{rxn} = q at constant pressure = -1234.7 kJ/mol rxn.

$$\Delta E = q + w$$
$$= (-1234.7 \ kJ) + (-4.957 \ kJ)$$
$$= -1239.7 \ kJ$$

INSIGHT:	Notice that ΔH and ΔE are almost the same value. They differ only by the amount of work the system does, -4.957 kJ. This agrees with the definition of enthalpy change, $\Delta H = \Delta E + P\Delta V = \Delta E + \Delta n_{gas}RT$ at constant temperature and pressure.

Reaction Enthalpy Change Calculation

7. *What is the enthalpy change for this reaction at standard conditions?*

$$C_2H_5OH(l) \ + \ 3 \ O_2(g) \rightarrow 2 \ CO_2(g) \ + \ 3 \ H_2O(g)$$

The correct answer is: ΔH^0_{rxn} = -1234.7 kJ/mol rxn

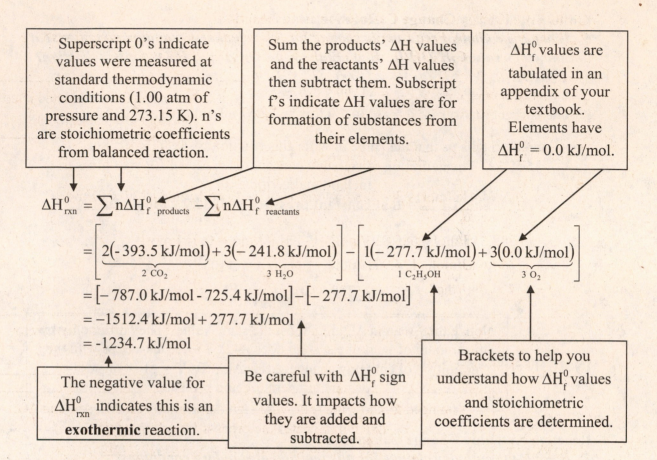

Superscript 0's indicate values were measured at standard thermodynamic conditions (1.00 atm of pressure and 273.15 K). n's are stoichiometric coefficients from balanced reaction.

Sum the products' ΔH values and the reactants' ΔH values then subtract them. Subscript f's indicate ΔH values are for formation of substances from their elements.

ΔH_f^0 values are tabulated in an appendix of your textbook. Elements have $\Delta H_f^0 = 0.0$ kJ/mol.

$$\Delta H_{rxn}^0 = \sum n\Delta H_f^0{}_{products} - \sum n\Delta H_f^0{}_{reactants}$$

$$= \left[\underbrace{2(-393.5\,\text{kJ/mol})}_{2\ CO_2} + \underbrace{3(-241.8\,\text{kJ/mol})}_{3\ H_2O}\right] - \left[\underbrace{1(-277.7\,\text{kJ/mol})}_{1\ C_2H_5OH} + \underbrace{3(0.0\,\text{kJ/mol})}_{3\ O_2}\right]$$

$$= \left[-787.0\,\text{kJ/mol} - 725.4\,\text{kJ/mol}\right] - \left[-277.7\,\text{kJ/mol}\right]$$

$$= -1512.4\,\text{kJ/mol} + 277.7\,\text{kJ/mol}$$

$$= -1234.7\,\text{kJ/mol}$$

The negative value for ΔH_{rxn}^0 indicates this is an **exothermic** reaction.

Be careful with ΔH_f^0 sign values. It impacts how they are added and subtracted.

Brackets to help you understand how ΔH_f^0 values and stoichiometric coefficients are determined.

Entropy Change Calculation for a Reaction
8. *What is the entropy change for this reaction at standard conditions?*
$$C_2H_5OH(l) + 3\,O_2(g) \rightarrow 2\,CO_2(g) + 3\,H_2O(g)$$

The correct answer is: $\Delta S^0{}_{rxn} = 217.3$ J/mol

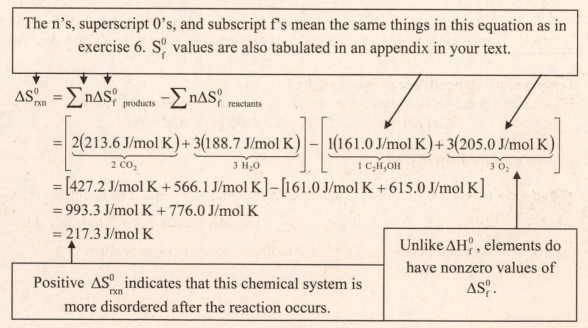

The n's, superscript 0's, and subscript f's mean the same things in this equation as in exercise 6. S_f^0 values are also tabulated in an appendix in your text.

$$\Delta S_{rxn}^0 = \sum n\Delta S_f^0{}_{products} - \sum n\Delta S_f^0{}_{reactants}$$

$$= \left[\underbrace{2(213.6\,\text{J/mol K})}_{2\ CO_2} + \underbrace{3(188.7\,\text{J/mol K})}_{3\ H_2O}\right] - \left[\underbrace{1(161.0\,\text{J/mol K})}_{1\ C_2H_5OH} + \underbrace{3(205.0\,\text{J/mol K})}_{3\ O_2}\right]$$

$$= \left[427.2\,\text{J/mol K} + 566.1\,\text{J/mol K}\right] - \left[161.0\,\text{J/mol K} + 615.0\,\text{J/mol K}\right]$$

$$= 993.3\,\text{J/mol K} + 776.0\,\text{J/mol K}$$

$$= 217.3\,\text{J/mol K}$$

Positive ΔS_{rxn}^0 indicates that this chemical system is more disordered after the reaction occurs.

Unlike ΔH_f^0, elements do have nonzero values of ΔS_f^0.

Gibbs Free Energy Change Calculation for a Reaction

9. *What is the Gibbs Free Energy change for this reaction at standard conditions?*
$$C_2H_5OH(l) \ + \ 3 \, O_2(g) \rightarrow 2 \, CO_2(g) \ + \ 3 \, H_2O(g)$$

The correct answer is: ΔG^0_{rxn} = -1299.7 kJ/mol

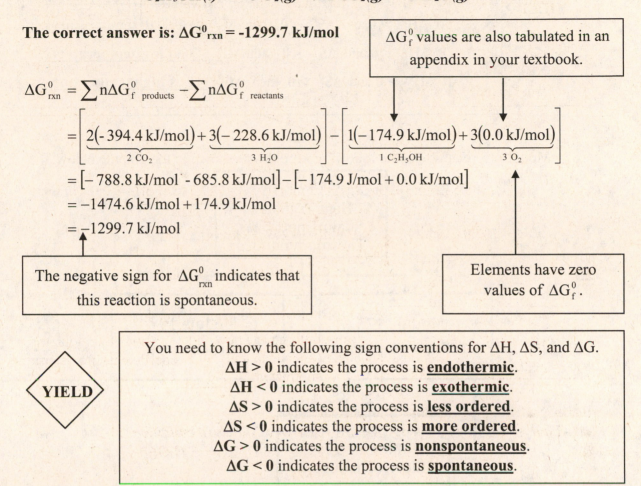

ΔG_f^0 values are also tabulated in an appendix in your textbook.

$$\Delta G^0_{rxn} = \sum n\Delta G^0_{f \ products} - \sum n\Delta G^0_{f \ reactants}$$

$$= \left[\underbrace{2(-394.4 \, kJ/mol)}_{2 \ CO_2} + \underbrace{3(-228.6 \, kJ/mol)}_{3 \ H_2O} \right] - \left[\underbrace{1(-174.9 \, kJ/mol)}_{1 \ C_2H_5OH} + \underbrace{3(0.0 \, kJ/mol)}_{3 \ O_2} \right]$$

$$= \left[-788.8 \, kJ/mol - 685.8 \, kJ/mol \right] - \left[-174.9 \, J/mol + 0.0 \, kJ/mol \right]$$

$$= -1474.6 \, kJ/mol + 174.9 \, kJ/mol$$

$$= -1299.7 \, kJ/mol$$

The negative sign for ΔG^0_{rxn} indicates that this reaction is spontaneous.

Elements have zero values of ΔG_f^0.

YIELD

You need to know the following sign conventions for ΔH, ΔS, and ΔG.
$\Delta H > 0$ indicates the process is **endothermic**.
$\Delta H < 0$ indicates the process is **exothermic**.
$\Delta S > 0$ indicates the process is **less ordered**.
$\Delta S < 0$ indicates the process is **more ordered**.
$\Delta G > 0$ indicates the process is **nonspontaneous**.
$\Delta G < 0$ indicates the process is **spontaneous**.

 TIP   Values for ΔH_f°, S_f°, and ΔG_f° are found in your textbook appendix. Of the three, only S_f° for a substance in its elemental state may have a non-zero value.

Temperature Dependence of Spontaneity

10. *Can this reaction become nonspontaneous if the temperature is changed?*
$$C_2H_5OH(l) \ + \ 3 \, O_2(g) \rightarrow 2 \, CO_2(g) \ + \ 3 \, H_2O(g)$$

   The correct answer is no.

From exercises 6 and 7 this reaction has $\Delta H^0_{rxn} < 0$ and $\Delta S^0_{rxn} > 0$. The definition of ΔG^0_{rxn} is $\Delta G^0_{rxn} = \Delta H^0_{rxn} - T \Delta S^0_{rxn}$. In this case, ΔG^0_{rxn} = (negative quantity) – T (positive quantity) is $\Delta G^0_{rxn} < 0$ at all temperatures. ($\Delta G^0_{rxn} < 0$ indicates a spontaneous reaction.)

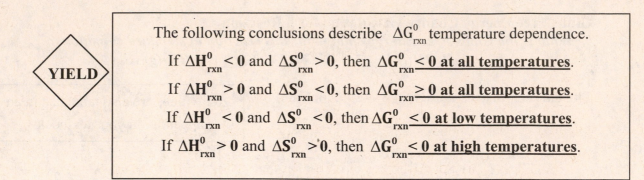

The following conclusions describe ΔG^0_{rxn} temperature dependence.

If $\Delta H^0_{rxn} < 0$ and $\Delta S^0_{rxn} > 0$, then ΔG^0_{rxn} **< 0 at all temperatures**.

If $\Delta H^0_{rxn} > 0$ and $\Delta S^0_{rxn} < 0$, then ΔG^0_{rxn} **> 0 at all temperatures**.

If $\Delta H^0_{rxn} < 0$ and $\Delta S^0_{rxn} < 0$, then ΔG^0_{rxn} **< 0 at low temperatures**.

If $\Delta H^0_{rxn} > 0$ and $\Delta S^0_{rxn} > 0$, then ΔG^0_{rxn} **< 0 at high temperatures**.

Module 15 relates to some following Modules as shown in the graphic below.

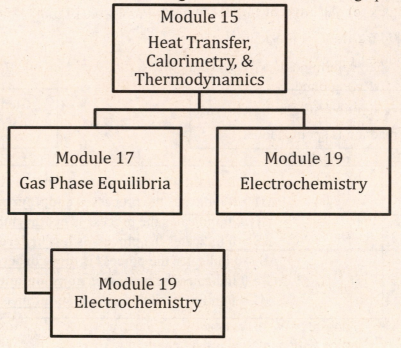

Practice Test Four
Modules 12-15

Level 1   1. List six strong acids and seven strong bases that are water soluble.

Level 1   2. Choose all of the true statements from this list:

   a) All Arrhenius bases are also Brønsted-Lowry bases.
   b) All Brønsted-Lowry bases are also Arrhenius bases.
   c) All Arrhenius acids are also Lewis acids.
   d) All Lewis acids are also Arrhenius acids.
   e) All Arrhenius acids are also Brønsted-Lowry acids *and* Lewis acids.

Level 3   3. Determine the *strongest* intermolecular force possible for each compound.
   a) CH_4
   b) CH_2Cl_2
   c) CH_3COOH
   d) HF
   e) PCl_3

Level 3   4. Arrange the following in order of *decreasing* boiling point:
   CaO, CCl_4, CH_2Br_2, CH_3COOH

Level 1   5. What volume is occupied by a 42.5 g sample of CH_4 gas at 1.34 atm and 32°C?

Level 2   6. Tungsten (W) has a density of 19.3 g/cm^3 and crystallizes in a cubic lattice with a unit cell edge length of 3.16×10^{-10} m. Which cubic unit cell is present in tungsten crystals?

Level 3   7. Determine the %w/w and the $X_{phosphoric\ acid}$ of a 2.75 *m* aqueous H_3PO_4 solution.

Level 1   8. The freezing point of an aqueous solution containing 15.0 g of a nonelectrolyte in 150. mL of water is -5.40 °C. What is the molecular weight of the nonelectrolyte? K_f for water is 1.86 °C/m.

Level 1   9. Calculate the amount of heat emitted to convert 21.3 grams of steam at 230.0°C to ice at -12.6°C. Necessary specific heats are: $H_2O(s)$ = 2.09 J/g·°C; $H_2O(l)$ = 4.18 J/g·°C; $H_2O(g)$ = 2.03 J/g·°C. Heats of fusion and vaporization are: $H_2O(s)$ = 333 J/g; $H_2O(l)$ = 2260 J/g, respectively.

Level 1   10. Using the table of thermodynamic data provided, calculate ΔH^0_{rxn} for this reaction. Is the reaction endothermic or exothermic?

$$SiH_4(g) + 2\ O_2(g) \rightarrow SiO_2(s) + 2\ H_2O(l)$$

Species	ΔH^0_f (kJ/mol)
$SiH_4(g)$	34.3
$SiO_2(s)$	-910.9
$H_2O(l)$	-285.8

Module 16 Predictor Questions

The following questions may help you determine the extent you need to study this module. Questions are ranked according to ability.

      Level 1 = basic proficiency
      Level 2 = mid level proficiency
      Level 3 = high proficiency

If you can correctly answer Level 3 questions you probably do not need to spend much time on this module. If you can only answer Level 1 problems, you should review this module.

Level 2   1. How are the rates of the disappearance of O_3 and the appearance of C_2H_4O and O_2 related to the disappearance of C_2H_4 in this reaction?
$$C_2H_4(g) + O_3(g) \rightarrow C_2H_4O(g) + O_2(g)$$

Level 1   2. Determine the rate-law expression for this reaction using the provided data.
$$2\,A + B_2 + C \rightarrow A_2B + BC$$

Trial	Initial [A]	Initial [B₂]	Initial [C]	Initial rate of formation of BC
1	0.20 M	0.20 M	0.20 M	2.4 x 10^{-6} $M \cdot min^{-1}$
2	0.40 M	0.30 M	0.20 M	9.6 x 10^{-6} $M \cdot min^{-1}$
3	0.20 M	0.30 M	0.20 M	2.4 x 10^{-6} $M \cdot min^{-1}$
4	0.20 M	0.40 M	0.40 M	4.8 x 10^{-6} $M \cdot min^{-1}$

Level 1   3. The second order reaction $2\,CH_4 \rightarrow C_2H_2 + 3\,H_2$ has a rate constant of 5.76 $M^{-1} \cdot min^{-1}$ at 1600 K. How long, in min, will it take for the concentration of CH_4 to reduce from 0.89 M to 5.25 x 10^{-4} M?

Level 1   4. Consider the following rate law expression: rate = $k[A]^2[B]$. Which is **not** true about a reaction having this expression?
a) The reaction is first order in B.
b) The reaction is overall third order.
c) The reaction is second order in A.
d) A and B must both be reactants.
e) Doubling the concentration of A doubles the rate.

Level 1   5. This reaction is first order with respect to CS_2 with $k = 2.8$ x 10^{-7} s^{-1} at 1000°C.
$$CS_2 \rightarrow CS + S$$
If the initial CS_2 concentration is 2.0 M, what is the CS_2 concentration 28 days later?

Level 1   6. Consider again the decomposition reaction of CS_2 (use the same k

value given above). How many days must pass before a 10.0 g sample of CS_2 decomposes so that only 2.00 g of CS_2 remains?

Level 1   7. What is the half-life, in days, of the CS_2 decomposition reaction described in question 6?

Level 1   8. This reaction is second order with $k = 0.0442\ M^{-1}s^{-1}$.

$$2\ C_2F_4 \rightarrow C_4F_8$$

If the initial C_2F_4 concentration is 0.0675 M, what is the concentration of C_2F_4 100. seconds after the reaction begins? What is the half-life of this reaction at these conditions?

Level 2   9. Find E_a for a reaction where the rate constant quadruples as the temperature increases from 298 K to 318 K.

Module 16 Predictor Question Solutions

1. How are the rates of the disappearance of O_3 and the appearance of C_2H_4O and O_2 related to the disappearance of C_2H_4 in this reaction?

$$C_2H_4(g) + O_3(g) \rightarrow C_2H_4O\,(g) + O_2(g)$$

$$\frac{-\Delta[C_2H_4]}{\Delta t} = \frac{-\Delta[O_3]}{\Delta t} = \frac{+\Delta[C_2H_4O]}{\Delta t} = \frac{+\Delta[O_2]}{\Delta t}$$

2. Determine the rate-law expression for this reaction using the experimental data provided.

$$2\,A + B_2 + C \rightarrow A_2B + BC$$

Trial	Initial [A]	Initial [B₂]	Initial [C]	Initial rate of formation of BC
1	0.20 M	0.20 M	0.20 M	2.4 x 10^{-6} $M \cdot min^{-1}$
2	0.40 M	0.30 M	0.20 M	9.6 x 10^{-6} $M \cdot min^{-1}$
3	0.20 M	0.30 M	0.20 M	2.4 x 10^{-6} $M \cdot min^{-1}$
4	0.20 M	0.40 M	0.40 M	4.8 x 10^{-6} $M \cdot min^{-1}$

When [A] is doubled (trial 3 to trial 2) as [B] and [C] are held constant, the rate increases 4 times. We conclude the reaction is second order with respect to A.

When [B] is increased by one and a half (trial 1 to trial 3) as [A] and [C] are held constant, the rate does not change. We conclude the reaction is zero order with respect to B.

When [C] is doubled (trial 3 to trial 4) as [A] is held constant ([B] changes, but it has already been determined that this does not affect the rate), the rate doubles. We conclude the reaction is first order with respect to C.

The rate law expression is: rate = $k[A]^2[B]^0[C]^1$

3. The second order reaction $2\,CH_4 \rightarrow C_2H_2 + 3\,H_2$ has a rate constant of 5.76 $M^{-1} \cdot min^{-1}$ at 1600 K. How long, in min, will it take for the concentration of CH_4 to reduce from 0.89 M to 5.25 x 10^{-4} M?

$$\frac{1}{[A]} - \frac{1}{[A_0]} = kt$$

$$\frac{1}{5.24 \times 10^{-4}\,M} - \frac{1}{0.89\,M} = (5.76\,M^{-1}min^{-1})t$$

$$t = 330\,min$$

4. Consider the following rate law expression: rate = $k[A]^2[B]$. Which is **not** true about a reaction having this expression?

a) The reaction is first order in B.
b) The reaction is overall third order.
c) The reaction is second order in A.
d) A and B must both be reactants.
e) Doubling the concentration of A doubles the rate.

Statement e) is not true. The reaction is second order with respect to A, so doubling [A] increases the rate by a factor of 4 rather than doubling it.

5. This reaction is first order with respect to CS_2 with $k = 2.8 \times 10^{-7}$ s^{-1} at 1000°C.
$$CS_2 \rightarrow CS + S$$
If the initial CS_2 concentration is 2.0 M, what is the CS_2 concentration 28 days later?

$$(28\,days)\left(\frac{24\,h}{1\,day}\right)\left(\frac{60\,min}{1\,h}\right)\left(\frac{60\,sec}{1\,min}\right) = 2.42 \times 10^6 \, s$$

$$[A] = [A_0]e^{-kt}$$

$$[A] = (2.0\,M)e^{-(2.8 \times 10^{-7}\,s^{-1})(2.42 \times 10^6\,s)} = 1.02\,M$$

6. Consider again the decomposition reaction of CS_2 (use the same k value given above). How many days must pass before a 10.0 g sample of CS_2 decomposes so that only 2.00 g of CS_2 remains?

$$[A] = [A_0]e^{-kt} \Rightarrow \frac{\ln\dfrac{[A]}{[A_0]}}{-k} = t$$

$$t = \frac{\ln\dfrac{(2.00\,g)}{(10.0\,g)}}{-2.8 \times 10^{-7}\,s^{-1}} = 5.75 \times 10^6 \, s = 66.5\,days$$

7. What is the half-life, in days, of the CS_2 decomposition reaction described in question 6?

$$kt_{1/2} = 0.693 \Rightarrow t_{1/2} = \frac{0.693}{k}$$

$$t_{1/2} = \frac{0.693}{2.8 \times 10^{-7}\,s^{-1}} = 2.48 \times 10^6\,s$$

$$t_{1/2} = 28.6\,days$$

8. This reaction is second order with $k = 0.0442$ $M^{-1}s^{-1}$.
$$2\,C_2F_4 \rightarrow C_4F_8$$
If the initial C_2F_4 concentration is 0.0675 M, what is the concentration of C_2F_4 100. seconds after the reaction begins? What is the half-life of this reaction at these conditions?

172

$$\frac{1}{[A]} - \frac{1}{[A_0]} = kt$$

$$\frac{1}{[A]} = \frac{1}{[A_0]} + kt = \frac{1}{(0.0675\ M)} + (0.0442\ M^{-1}s^{-1})(100.s) = 19.2\ M^{-1}$$

$$[A] = 0.0520\ M$$

$$kt_{1/2} = \frac{1}{[A_0]}$$

$$(0.0442\ M^{-1}s^{-1})t_{1/2} = \frac{1}{(0.0675\ M)}$$

$$t_{1/2} = 335\ s$$

9. Find E_a for a reaction where the rate constant quadruples as the temperature increases from 298 K to 318 K.

$$\ln\frac{k_2}{k_1} = \frac{E_a}{R}\left(\frac{1}{T_1} - \frac{1}{T_2}\right)$$

$$\frac{\ln(4)}{\left(\frac{1}{298\ K} - \frac{1}{318\ K}\right)} = \frac{E_a}{8.314\ \dfrac{J}{mol \cdot K}}$$

$$E_a = 5.46 \times 10^4\ J\,/\,mol$$

Module 16
Chemical Kinetics

Introduction

Chemical kinetics is the study of reaction rates, how quickly or slowly a reaction occurs. Chemists can control several factors that change reaction rates including temperature and reactant concentrations. This module describes:

1. the rate relationships of one reactant to other reactants and products
2. how to determine the rate order of a reactant from experimental data
3. integrated rate laws for first and second order reactions
4. the effect of temperature on the rate of a reaction using the Arrhenius equation.

Module 16 Key Equations & Concepts

1. $\text{rate} \propto \dfrac{-\Delta[A]}{\Delta t} = \dfrac{-\Delta[B]}{b\Delta t} = \dfrac{+\Delta[C]}{c\Delta t}$ **for the reaction** $A + bB \rightarrow cC$

   This is the reaction rate definition based on the concentrations of reactants or products. [A] represents the molar concentration of substance A and similarly for [B] and [C]. Notice that reactant concentrations, A and B, decrease (-) with time, t, while product concentrations, C, increase with time (+).

2. $[A] = [A_0]e^{-kt}$

   This is the ***first order kinetics*** integrated rate law for chemical reactions. It is used to calculate *either* the reactant concentration some time after a reaction has started *or* the time required for a reactant concentration to reach a specified amount. [A] is the concentration after the time has passed, $[A_0]$ is the initial concentration, k is the rate constant, and t is the time.

3. $k\, t_{1/2} = 0.693$

   The ***first order reaction*** half-life relationship defines the half-life of a first order reaction given the rate constant or vice versa. A half-life is the time necessary for the initial concentration to become one-half of its initial amount.

4. $\dfrac{1}{[A]} - \dfrac{1}{[A_0]} = kt$

   This is the ***second order kinetics*** integrated rate law for chemical reactions. It is used to determine *either* the concentration of a reactant a certain amount of time after a reaction has started *or* the amount of time required for the concentration of a reactant to reach a specified amount.

5. $k\, t_{1/2} = \dfrac{1}{[A_0]}$

   The ***second order reaction*** half-life relationship defines the half-life of a second order reaction given the rate constant or vice versa.

---

6. $$\ln \frac{k_2}{k_1} = \frac{E_a}{R}\left(\frac{1}{T_1} - \frac{1}{T_2}\right)$$

The **Arrhenius equation** describes how a reaction rate of changes as the reaction temperature changes. E_a is the reaction activation energy; k_2 and k_1 are the rate constants at temperatures T_1 and T_2; R is the gas constant.

Sample Exercises
Reaction Rate Based on Product and Reactant Concentrations
1. *How are the rates of disappearance of O_2 and appearance of H_2O related to the rate of disappearance of H_2 in this reaction?*

$$2\ H_2(g)\ +\ O_2(g)\ \rightarrow\ 2\ H_2O(g)$$

The correct answer is $\dfrac{-\Delta[H_2]}{\Delta t} = \dfrac{-2\Delta[O_2]}{\Delta t} = \dfrac{+\Delta[H_2O]}{\Delta t}$.

Positive and negative signs indicate whether the concentrations are increasing (+) or decreasing (-) with time.

$$\frac{-\Delta[H_2]}{2\ \Delta t} = \frac{-\Delta[O_2]}{\Delta t} = \frac{+\Delta[H_2O]}{2\ \Delta t}$$

or

$$\frac{-\Delta[H_2]}{\Delta t} = \frac{-2\Delta[O_2]}{\Delta t} = \frac{+\Delta[H_2O]}{\Delta t}$$

This indicates that the H_2 concentration is decreasing at twice the rate O_2 is decreasing and the same rate H_2O is appearing.

Reaction Order Determination from Experimental Data
2. *The following experimental data were obtained for the chemical reaction:*

$$(C_2H_5)_3N + C_2H_5Br \rightarrow (C_2H_5)_4NBr$$

Experiment	$[(C_2H_5)_3N]$ (M)	$[C_2H_5Br]$ (M)	Relative rate (M/min)
1	0.10	0.10	3.0
2	0.20	0.10	6.0
3	0.10	0.30	9.0

What is the rate equation for this reaction and the value of the rate constant, k?
The correct answer is: rate = k $[(C_2H_5)_3N]^1$ $[C_2H_5Br]^1$ and k = 3.0 x 10^2 M^{-1} min^{-1}.

Problems of this type present a set of data in which the concentration of one reactant changes while the other reactant concentrations remain constant. For example, compare experiments 1 and 2. Notice that $[(C_2H_5)_3N]$ doubles, 0.10 M to 0.20 M, and the $[C_2H_5Br]$ remains constant at 0.10 M. Thus the concentration rate effect has been isolated to $[(C_2H_5)_3N]$. Now, look at the relative rates for experiments 1 and 2 which changes from 3.0 to 6.0 M/min, it doubles. From that information we conclude the reaction is 1st order with respect to $[(C_2H_5)_3N]$.

Next compare experiments 1 and 3 where $[(C_2H_5)_3N]$ remains constant at $0.10\ M$ and $[C_2H_5Br]$ triples, from $0.10\ M$ to $0.30\ M$. The rate also triples from 3.0 to 9.0 M/min. Consequently, this reaction is 1st order with respect to $[C_2H_5Br]$.

Both pieces of reaction data are required to write the rate law equation for this reaction.

$$\text{rate} = k\ [(C_2H_5)_3N]^1\ [C_2H_5Br]^1$$

Chemists say that this reaction is 1st order with respect to $[(C_2H_5)_3N]$, 1st order with respect to $[C_2H_5Br]$ and 2nd order overall. (Overall order is the sum of the individual orders.)

CAUTION

A very common mistake made by students is assuming the reaction order is determined by the stoichiometric coefficients in the balanced reaction. This is not correct. **The only method to determine the reaction order is experimental data analysis as done in this problem.**

The rate constant value, k, is determined using data from experiments 1, 2, or 3. For example let's choose experiment 3's data.

The rate values and concentrations of both $(C_2H_5)_3N$ and C_2H_5Br come from experiment 3.

Units for k of $1/(M \cdot$ time) are correct for 2nd order reactions.

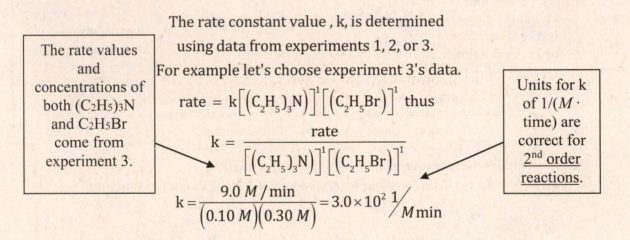

$$\text{rate} = k\left[\left(C_2H_5\right)_3N\right]^1\left[C_2H_5Br\right]^1 \text{ thus}$$

$$k = \frac{\text{rate}}{\left[\left(C_2H_5\right)_3N\right]^1\left[C_2H_5Br\right]^1}$$

$$k = \frac{9.0\ M/\text{min}}{\left(0.10\ M\right)\left(0.30\ M\right)} = 3.0\times10^2\ \frac{1}{M\,\text{min}}$$

First Order Integrated Rate Law

3. *The following reaction is first order with respect to $[NH_2NO_2]$ having a value for the rate constant, k, of $9.3 \times 10^{-5}\ s^{-1}$. If the initial $[NH_2NO_2] = 2.0\ M$, what is the $[NH_2NO_2]$ 30.0 minutes after the reaction starts?*

$$NH_2NO_2(aq) \ \rightarrow\ N_2O(g)\ +\ H_2O(l)$$

The correct answer is: 1.7 M.

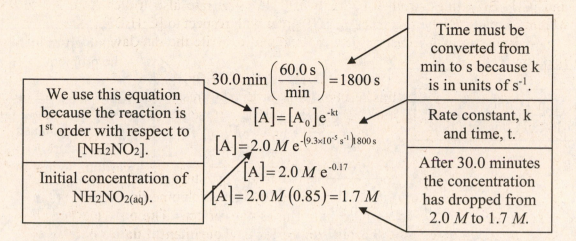

We use this equation because the reaction is 1st order with respect to [NH$_2$NO$_2$].	$30.0 \, min \left(\dfrac{60.0 \, s}{min} \right) = 1800 \, s$	Time must be converted from min to s because k is in units of s^{-1}.
	$[A] = [A_0] e^{-kt}$	Rate constant, k and time, t.
Initial concentration of NH$_2$NO$_{2(aq)}$.	$[A] = 2.0 \, M \, e^{-(9.3 \times 10^{-5} \, s^{-1})1800 \, s}$ $[A] = 2.0 \, M \, e^{-0.17}$ $[A] = 2.0 \, M \, (0.85) = 1.7 \, M$	After 30.0 minutes the concentration has dropped from 2.0 M to 1.7 M.

4. The following reaction is first order with respect to [NH$_2$NO$_2$] having a value for the rate constant, k, of 9.3 x 10^{-5} s^{-1}. If the initial [NH$_2$NO$_2$] = 2.0 M, how long will it take for the [NH$_2$NO$_2$] = 1.5 M?

$$NH_2NO_2(aq) \rightarrow N_2O(g) + H_2O(l)$$

The correct answer is: 3.1 x 10^3 s or 52 min.

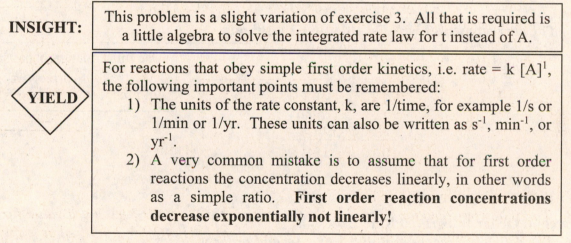

INSIGHT:	This problem is a slight variation of exercise 3. All that is required is a little algebra to solve the integrated rate law for t instead of A.
YIELD	For reactions that obey simple first order kinetics, i.e. rate = k [A]^1, the following important points must be remembered: 1) The units of the rate constant, k, are 1/time, for example 1/s or 1/min or 1/yr. These units can also be written as s^{-1}, min^{-1}, or yr^{-1}. 2) A very common mistake is to assume that for first order reactions the concentration decreases linearly, in other words as a simple ratio. **First order reaction concentrations decrease exponentially not linearly!**

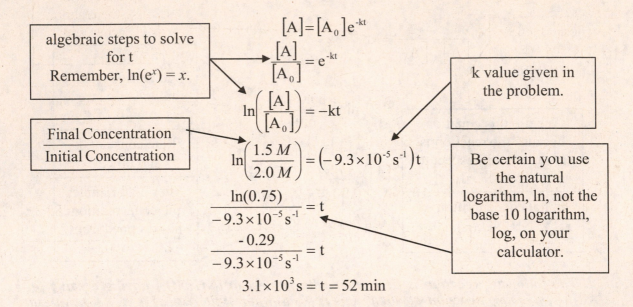

algebraic steps to solve for t
Remember, $\ln(e^x) = x$.

$$[A] = [A_0]e^{-kt}$$

$$\frac{[A]}{[A_0]} = e^{-kt}$$

k value given in the problem.

$$\ln\left(\frac{[A]}{[A_0]}\right) = -kt$$

Final Concentration / Initial Concentration

$$\ln\left(\frac{1.5\,M}{2.0\,M}\right) = \left(-9.3 \times 10^{-5}\,s^{-1}\right)t$$

Be certain you use the natural logarithm, ln, not the base 10 logarithm, log, on your calculator.

$$\frac{\ln(0.75)}{-9.3 \times 10^{-5}\,s^{-1}} = t$$

$$\frac{-0.29}{-9.3 \times 10^{-5}\,s^{-1}} = t$$

$$3.1 \times 10^3\,s = t = 52\,min$$

5. **The following reaction is first order with respect to [NH₂NO₂] having a value for the rate constant, k, of 9.3 x 10⁻⁵ s⁻¹. What is the half-life of this reaction?**

$$NH_2NO_2(aq) \rightarrow N_2O(g) + H_2O(l)$$

The correct answer is: 7.5 x 10³ s or 1.2 x 10² min.

$$kt_{1/2} = 0.693$$

Use this equation because the reaction is 1st order with respect to [NH₂NO₂].

$$t_{1/2} = \frac{0.693}{k}$$

$$t_{1/2} = \frac{0.693}{9.3 \times 10^{-5}\,s^{-1}}$$

$$t_{1/2} = 7.5 \times 10^3\,s = 1.2 \times 10^2\,min$$

Second Order Integrated Rate Law

6. **The following reaction at 400.0 K is second order with respect to [CF₃] having a value for the rate constant, k, of 2.51 x 10¹⁰ M⁻¹s⁻¹. If the initial [CF₃] = 2.0 M, what is the [CF₃] 4.25 x 10⁻¹⁰ seconds after the reaction has started?**

$$2\,CF_3(g) \rightarrow C_2F_6(g)$$

The correct answer is: 8.93 x 10⁻² M.

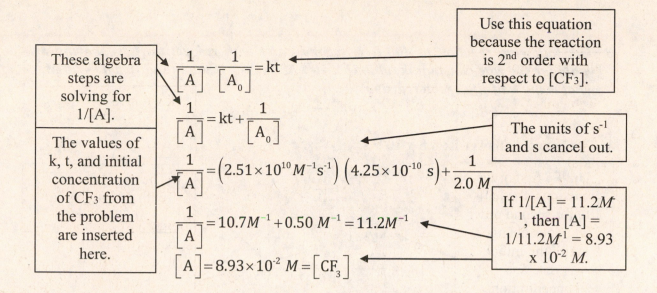

These algebra steps are solving for $1/[A]$.

The values of k, t, and initial concentration of CF_3 from the problem are inserted here.

Use this equation because the reaction is 2nd order with respect to $[CF_3]$.

$$\frac{1}{[A]} - \frac{1}{[A_0]} = kt$$

$$\frac{1}{[A]} = kt + \frac{1}{[A_0]}$$

$$\frac{1}{[A]} = \left(2.51 \times 10^{10} M^{-1} s^{-1}\right)\left(4.25 \times 10^{-10}\ s\right) + \frac{1}{2.0\ M}$$

$$\frac{1}{[A]} = 10.7 M^{-1} + 0.50\ M^{-1} = 11.2 M^{-1}$$

$$[A] = 8.93 \times 10^{-2}\ M = [CF_3]$$

The units of s^{-1} and s cancel out.

If $1/[A] = 11.2 M^{-1}$, then $[A] = 1/11.2 M^{-1} = 8.93 \times 10^{-2}\ M$.

7. *The following reaction at 400.0 K is second order with respect to $[CF_3]$ having a value for the rate constant, k, of $2.51 \times 10^{10}\ M^{-1}s^{-1}$. If the initial $[CF_3] = 2.0\ M$, how long will it take for the $[CF_3] = 1.5\ M$?*

$$2\ CF_3(g) \rightarrow C_2F_6(g)$$

The correct answer is: 6.8×10^{-12} s.

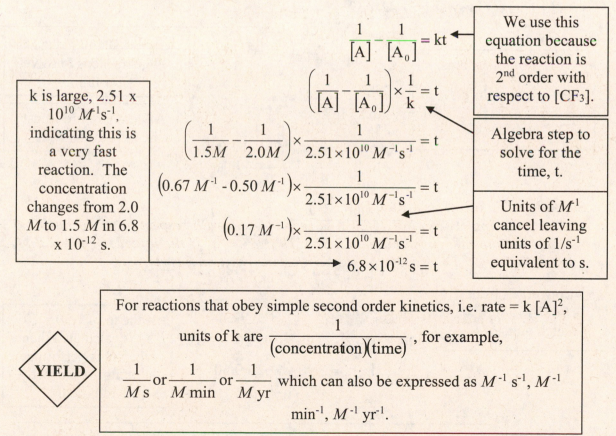

k is large, $2.51 \times 10^{10}\ M^{-1}s^{-1}$, indicating this is a very fast reaction. The concentration changes from 2.0 M to 1.5 M in 6.8×10^{-12} s.

We use this equation because the reaction is 2nd order with respect to $[CF_3]$.

$$\frac{1}{[A]} - \frac{1}{[A_0]} = kt$$

$$\left(\frac{1}{[A]} - \frac{1}{[A_0]}\right) \times \frac{1}{k} = t$$

$$\left(\frac{1}{1.5M} - \frac{1}{2.0M}\right) \times \frac{1}{2.51 \times 10^{10}\ M^{-1}s^{-1}} = t$$

$$\left(0.67\ M^{-1} - 0.50\ M^{-1}\right) \times \frac{1}{2.51 \times 10^{10}\ M^{-1}s^{-1}} = t$$

$$\left(0.17\ M^{-1}\right) \times \frac{1}{2.51 \times 10^{10}\ M^{-1}s^{-1}} = t$$

$$6.8 \times 10^{-12} s = t$$

Algebra step to solve for the time, t.

Units of M^{-1} cancel leaving units of $1/s^{-1}$ equivalent to s.

YIELD

For reactions that obey simple second order kinetics, i.e. rate = k $[A]^2$, units of k are $\dfrac{1}{(\text{concentration})(\text{time})}$, for example, $\dfrac{1}{M\ s}$ or $\dfrac{1}{M\ \text{min}}$ or $\dfrac{1}{M\ \text{yr}}$ which can also be expressed as $M^{-1}\ s^{-1}$, $M^{-1}\ \text{min}^{-1}$, $M^{-1}\ \text{yr}^{-1}$.

8. **The following reaction at 400 K is second order with respect to [CF₃] having a value for the rate constant, k, of 2.51 x 10¹⁰ M⁻¹s⁻¹. If the initial [CF₃] = 2.0 M, what is the half-life of the reaction?**

$$2\ CF_3(g) \rightarrow C_2F_6(g)$$

The correct answer is: 6.8×10^{-12} s.

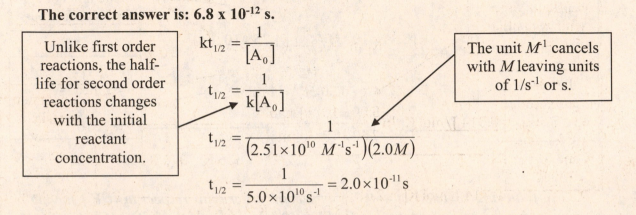

Unlike first order reactions, the half-life for second order reactions changes with the initial reactant concentration.

$$kt_{1/2} = \frac{1}{[A_0]}$$

$$t_{1/2} = \frac{1}{k[A_0]}$$

The unit M^{-1} cancels with M leaving units of $1/s^{-1}$ or s.

$$t_{1/2} = \frac{1}{(2.51 \times 10^{10}\ M^{-1}s^{-1})(2.0M)}$$

$$t_{1/2} = \frac{1}{5.0 \times 10^{10}\ s^{-1}} = 2.0 \times 10^{-11}\ s$$

Arrhenius Equation

9. **A reaction has an activation energy of 52.0 kJ/mol and a rate constant, k, of 7.50 x 10² s⁻¹ at 300.0 K. What is the rate constant for this reaction at 350.0 K?**

The correct answer is: 1.42×10^4 s⁻¹.

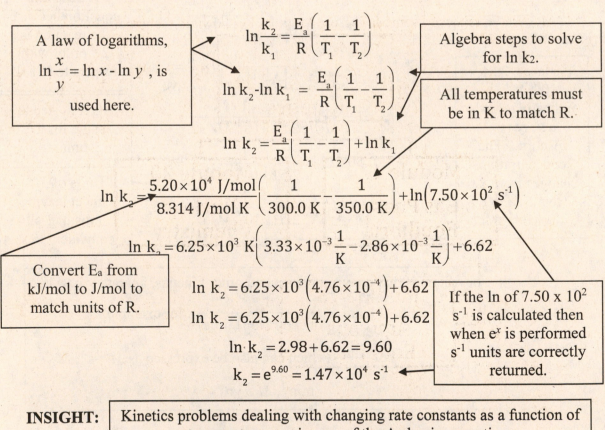

A law of logarithms, $\ln \dfrac{x}{y} = \ln x - \ln y$, is used here.

$$\ln \frac{k_2}{k_1} = \frac{E_a}{R}\left(\frac{1}{T_1} - \frac{1}{T_2}\right)$$

Algebra steps to solve for ln k₂.

$$\ln k_2 - \ln k_1 = \frac{E_a}{R}\left(\frac{1}{T_1} - \frac{1}{T_2}\right)$$

All temperatures must be in K to match R.

$$\ln k_2 = \frac{E_a}{R}\left(\frac{1}{T_1} - \frac{1}{T_2}\right) + \ln k_1$$

$$\ln k_2 = \frac{5.20 \times 10^4\ J/mol}{8.314\ J/mol\ K}\left(\frac{1}{300.0\ K} - \frac{1}{350.0\ K}\right) + \ln\left(7.50 \times 10^2\ s^{-1}\right)$$

$$\ln k_2 = 6.25 \times 10^3\ K\left(3.33 \times 10^{-3}\frac{1}{K} - 2.86 \times 10^{-3}\frac{1}{K}\right) + 6.62$$

Convert E_a from kJ/mol to J/mol to match units of R.

$$\ln k_2 = 6.25 \times 10^3\left(4.76 \times 10^{-4}\right) + 6.62$$

$$\ln k_2 = 6.25 \times 10^3\left(4.76 \times 10^{-4}\right) + 6.62$$

$$\ln k_2 = 2.98 + 6.62 = 9.60$$

$$k_2 = e^{9.60} = 1.47 \times 10^4\ s^{-1}$$

If the ln of 7.50 x 10² s⁻¹ is calculated then when e^x is performed s⁻¹ units are correctly returned.

INSIGHT: Kinetics problems dealing with changing rate constants as a function of temperature require use of the Arrhenius equation.

180

10. **What is the activation energy of a reaction having a rate constant of 2.50 x 10² kJ/mol at 325K and a rate constant of 5.00 x 10² kJ/mol at 375 K?**
The correct answer is: 14.0 kJ/mol.

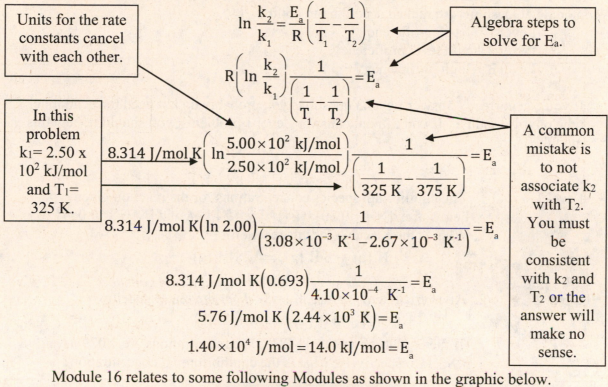

Units for the rate constants cancel with each other.

Algebra steps to solve for E_a.

$$\ln \frac{k_2}{k_1} = \frac{E_a}{R}\left(\frac{1}{T_1} - \frac{1}{T_2}\right)$$

$$R\left(\ln \frac{k_2}{k_1}\right)\frac{1}{\left(\frac{1}{T_1} - \frac{1}{T_2}\right)} = E_a$$

In this problem $k_1 = 2.50 \times 10^2$ kJ/mol and $T_1 = 325$ K.

A common mistake is to not associate k_2 with T_2. You must be consistent with k_2 and T_2 or the answer will make no sense.

$$8.314 \text{ J/mol K}\left(\ln \frac{5.00 \times 10^2 \text{ kJ/mol}}{2.50 \times 10^2 \text{ kJ/mol}}\right)\frac{1}{\left(\frac{1}{325 \text{ K}} - \frac{1}{375 \text{ K}}\right)} = E_a$$

$$8.314 \text{ J/mol K}(\ln 2.00)\frac{1}{\left(3.08 \times 10^{-3} \text{ K}^{-1} - 2.67 \times 10^{-3} \text{ K}^{-1}\right)} = E_a$$

$$8.314 \text{ J/mol K}(0.693)\frac{1}{4.10 \times 10^{-4} \text{ K}^{-1}} = E_a$$

$$5.76 \text{ J/mol K}\left(2.44 \times 10^3 \text{ K}\right) = E_a$$

$$1.40 \times 10^4 \text{ J/mol} = 14.0 \text{ kJ/mol} = E_a$$

Module 16 relates to some following Modules as shown in the graphic below.

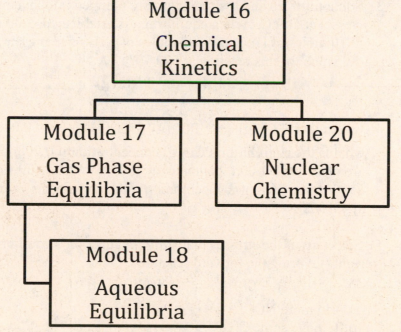

Module 16
Chemical Kinetics

Module 17
Gas Phase Equilibria

Module 20
Nuclear Chemistry

Module 18
Aqueous Equilibria

Module 17 Predictor Questions

The following questions may help you determine the extent you need to study this module. Questions are ranked according to ability.

Level 1 = basic proficiency
Level 2 = mid level proficiency
Level 3 = high proficiency

If you can correctly answer Level 3 questions you probably do not need to spend much time on this module. If you can only answer Level 1 problems, you should review this module.

Level 1   1. The equilibrium species concentrations for the reaction given below are: $[N_2] = 0.301$ M, $[H_2] = 0.240$ M, and $[NH_3] = 0.0541$ M. What is the K_c value for this reaction?

$$N_2(g) + 3 H_2(g) \rightleftharpoons 2 NH_3(g)$$

Level 2   2. At a certain temperature, $K_c = 14.5$ for the reaction below:

$$CO(g) + 2 H_2(g) \rightleftharpoons CH_3OH(g)$$

If the equilibrium CO and CH_3OH concentrations are 1.029 M and 1.86 M, respectively, what is the equilibrium H_2 concentration?

Level 1   3. Some nitrogen and hydrogen gas are pumped into an empty 5.00 L vessel at 500°C. When equilibrium is established, 3.00 moles of N_2, 2.10 moles of H_2, and 0.298 moles of NH_3 were present. What is the K_c value for this reaction at 500 °C?

$$N_2(g) + 3 H_2(g) \rightleftharpoons 2 NH_3(g)$$

Level 3   4. Given that: $PCl_5(g) \rightleftharpoons PCl_3(g) + Cl_2(g)$ has $K_c = 0.040$ at 450°C, what is the equilibrium PCl_5 (g) concentration if 0.20 mol of PCl_5 (g) are placed in a 1.00 L container at 450°C? What is the new equilibrium PCl_5 (g) concentration if the container's volume is halved at 450°C?

Level 1   5. Select all of the stresses that shift this reaction's equilibrium to the right (favoring the forward reaction).

$$2 NO(g) + Cl_2(g) \rightleftharpoons 2 NOCl(g) \ \Delta H < 0$$

a) Add more NOCl.
b) Remove some Cl_2.
c) Lower the temperature.
d) Add more NO.

Level 2   6. Consider the reaction below. Determine which reaction conditions
             will produce the maximum product yield.
$$A(g) + B(g) \rightleftharpoons D(g) + Heat$$
             a) 100°C, 50 atm
             b) 100 °C, 10 atm
             c) 500 °C, 50 atm, catalyst
             d) 100°C, 50 atm, catalyst
             e) 500°C, 10 atm, catalyst

Level 1   7. The equilibrium constant, K_c, for the following reaction is 0.0154 at
             high temperature. A mixture in a container at this high temperature
             has the following concentrations: $[H_2] = 1.11\ M$, $[I_2] = 1.30\ M$, $[HI] =$
             $0.181\ M$. Which of these statements concerning the reaction and the
             reaction quotient, Q, is true?
$$H_2(g) + I_2(g) \rightleftharpoons 2\,HI(g)$$
             a) $Q = K_c$
             b) $Q > K_c$; more HI will be produced
             c) $Q > K_c$; more H_2 and I_2 will be produced
             d) $Q < K_c$; more HI will be produced
             e) $Q < K_c$; more H_2 and I_2 will be produced

Level 3   8. Calculate the thermodynamic equilibrium constant at 25°C for a
             reaction having $\Delta G^0 = 11.3$ kJ per mol of reaction.
$$R = 8.314\ J/mol \cdot K$$

Module 17 Predictor Question Solutions

1. Species equilibrium concentrations for this reaction are: $[N_2] = 0.301$ M, $[H_2] = 0.240$ M, and $[NH_3] = 0.0541$ M. What is the value of K_c for this reaction?

$$N_2(g) + 3\,H_2(g) \rightleftharpoons 2\,NH_3(g)$$

$$K_c = \frac{[NH_3]^2}{[H_2]^3[N_2]} = \frac{[0.0541]^2}{[0.240]^3[0.301]} = 0.703$$

2. At a certain temperature, $K_c = 14.5$ for this reaction:

$$CO(g) + 2\,H_2(g) \rightleftharpoons CH_3OH(g)$$

If the CO and CH_3OH equilibrium concentrations are 1.029 M and 1.86 M, respectively, what is the H_2 equilibrium concentration?

$$K_c = \frac{\left[CH_3OH\right]}{\left[CO\right]\left[H_2\right]^2}$$

$$K_c = \frac{1.86\,M}{1.029\,M\left[H_2\right]^2} = 14.5$$

$$\left[H_2\right] = 0.353\,M$$

3. Some nitrogen and hydrogen gas are pumped into an empty 5.00 L vessel at 500°C. When equilibrium is established, 3.00 moles of N_2, 2.10 moles of H_2, and 0.298 moles of NH_3 were present. What is the K_c value for this reaction at 500 °C?

$$N_2(g) + 3\,H_2(g) \rightleftharpoons 2\,NH_3(g)$$

$$[NH_3] = \frac{0.298\ \text{mol}}{5.00\ \text{L}} = 0.0596\ M$$

$$[N_2] = \frac{3.00\ \text{mol}}{5.00\ \text{L}} = 0.600\ M$$

$$[H_2] = \frac{2.10\ \text{mol}}{5.00\ \text{L}} = 0.420\ M$$

$$K_p = \frac{[NH_3]^2}{[H_2]^3[N_2]} = \frac{[0.0596]^2}{[0.420]^3[0.600]} = 0.0799$$

4. Given that: $PCl_5(g) \rightleftharpoons PCl_3(g) + Cl_2(g)$ has $K_c = 0.040$ at 450°C, what is the equilibrium PCl_5 (g) concentration if 0.20 mol of PCl_5 (g) are placed in a 1.00 L container at 450°C? What is the new equilibrium PCl_5 (g) concentration if the container's volume is halved at 450°C?

$$PCl_5(g) \rightleftharpoons PCl_3(g) + Cl_2(g)$$

[Initial]	0.20 M	0	0
$\Delta[\ ]$	-x	+x	+x
[Equilibrium]	0.20 M - x	x	x

$$K_c = 0.040 = \frac{x^2}{0.20 - x}$$

$$x^2 + 0.040x - 0.008 = 0$$

Solve for x using the quadratic equation : x = 0.0717

$$[PCl_5] = 0.20\ M - 0.0717\ M = 0.128\ M$$

If the volume is halved, the initial concentration of PCl$_5$ doubles.

$$PCl_5(g) \rightleftharpoons PCl_3(g) + Cl_2(g)$$

[Initial]	0.40 M	0	0
$\Delta[\ ]$	-x	+x	+x
[Equilibrium]	0.40 M - x	x	x

$$K_c = 0.040 = \frac{x^2}{0.40 - x}$$

$$x^2 + 0.040x - 0.016 = 0$$

Solve for x using the quadratic equation : x = 0.108

$$[PCl_5] = 0.40\ M - 0.108\ M = 0.292\ M$$

5. Select all of the stresses that shift this reaction's equilibrium to the right (favoring the forward reaction).

$$2\ NO(g) + Cl_2(g) \rightleftharpoons 2\ NOCl(g)\ \ \Delta H < 0$$

a) Add more NOCl.
b) Remove some Cl$_2$.
c) Lower the temperature.
d) Add more NO.

The correct answers are c) and d).

The reaction is exothermic ($\Delta H < 0$). Lowering the temperature removes heat (which can be thought of as a product) driving the reaction to the right or product side. Adding more reactant (NO) also drives the reaction to right or product side.

6. Consider the reaction below. Determine which reaction conditions will produce the maximum product yield.

$$A(g) + B(g) \rightleftharpoons D(g) + Heat$$

a) 100°C, 50 atm

b) 100 °C, 10 atm

c) 500 °C, 50 atm, catalyst

d) 100°C, 50 atm, catalyst

e) 500°C, 10 atm, catalyst

The correct answers are a) and d).

Catalysts change rates, but they do not alter the position of equilibrium. The presence of a catalyst has no effect on product formation other than attaining equilibrium more quickly. Heat is a product in this exothermic reaction, so increasing temperature is equivalent to adding a product. This shifts equilibrium toward the reactants. Lower temperature is more favorable to product formation. This reaction occurs entirely in the gas phase. There are more moles of gas on the reactant side of the reaction, so an increase in pressure shifts the equilibrium to the product side.

7. The equilibrium constant, K_c, for the following reaction is 0.0154 at high temperature. A mixture in a container at this high temperature has the following concentrations: $[H_2] = 1.11$ M, $[I_2] = 1.30$ M, $[HI] = 0.181$ M. Which of these statements concerning the reaction and the reaction quotient, Q, is true?

$$H_2(g) + I_2(g) \rightleftharpoons 2\,HI(g)$$

a) $Q = K_c$

b) $Q > K_c$; more HI will be produced

c) $Q > K_c$; more H_2 and I_2 will be produced

d) $Q < K_c$; more HI will be produced

e) $Q < K_c$; more H_2 and I_2 will be produced

$$Q = \frac{[HI]^2}{[I_2][H_2]} = \frac{[0.181]^2}{[1.30][1.11]} = 0.0227$$

Since Q > K, the reaction is reactant favored. The correct answer is c).

8. Calculate the thermodynamic equilibrium constant at 25°C for a reaction having ΔG^0 = 11.3 kJ/mol of reaction.

$$R = 8.314 \text{ J/mol·K}$$

$$\Delta G^0_{rxn} = -RT \ln K$$

$$T = 25°C + 273.15 = 298.K$$

$$11.3\,kJ/mol = 11.3 \times 10^3\,J/mol$$

$$\ln K = -\frac{\Delta G^0_{rxn}}{RT} = -\frac{11.3 \times 10^3\,J/mol}{\left(8.314\,J/molK\right)298.K} = 4.56$$

$$K = e^{4.56} = 95.6$$

Module 17
Gas Phase Equilibria

Introduction

This module describes some basic calculations required for gas phase equilibria problems. Many ideas introduced here will be used again in Module 18 with slight modifications. This module describes how to:

1. determine the value of K_c, the equilibrium constant, and use it to predict if a reaction is product or reactant favored
2. calculate the species concentrations in a reaction
3. calculate K_p, the equilibrium constant in terms of the partial pressures of the gases, and its relation to K_c
4. use Le Châtelier's Principle and the reaction quotient, Q, to predict effects of temperature, pressure, and concentration changes on an equilibrium
5. examine the relationship of ΔG and K and calculate K value at different temperatures.

Module 17 Key Equations & Concepts

These equations refer to the generic reaction $aA + bB \rightleftharpoons cC + dD$ where a, b, c, and d are the stoichiometric coefficients for the reaction.

1. **The equilibrium constant, $K_c = \dfrac{[C]^c [D]^d}{[A]^a [B]^b}$**

   This equation determines whether the reaction is product or reactant favored (i.e. yields more reactants or products) as well as the reactant and product equilibrium concentrations. *Equilibrium concentrations are used in K_c.*

2. **The equilibrium constant for gas phase reactions, $K_p = \dfrac{\left(P_C\right)^c \left(P_D\right)^d}{\left(P_A\right)^a \left(P_B\right)^b}$**

   The equilibrium constant in terms of the partial pressures of the gases, K_P, serves the same purpose as K_c except gas partial pressures are used instead of equilibrium concentrations. K_p is used when it is easier to measure gas pressures instead of equilibrium concentrations.

3. **The reaction quotient, $Q = \dfrac{[C]^c [D]^d}{[A]^a [B]^b}$**

   Q, the reaction quotient has *the same form as K_c but uses nonequilibrium concentrations*. Q determines how the equilibrium position shifts for a system not at equilibrium to attain equilibrium.

4. $K_P = K_c (RT)^{\Delta n}$ **where**

   $\Delta n = \sum \textbf{moles of gaseous products} - \sum \textbf{moles of gaseous reactants}$

   This is the relationship of the partial pressure equilibrium constant, K_p, and the equilibrium constant, K_c.

---

5. $\Delta G^0_{rxn} = -RT \ln K$

K, the thermodynamic equilibrium constant, is related to the standard Gibbs Free Energy change with this relationship.

6. $\ln\left(\dfrac{K_{T_2}}{K_{T_1}}\right) = \dfrac{\Delta H^0}{R}\left(\dfrac{1}{T_1} - \dfrac{1}{T_2}\right)$

The van't Hoff equation defines equilibrium constant values at different temperatures.

Sample Exercises

Equilibrium Constant Expression Use

1. For this reaction at 298 K, equilibrium concentrations are [H₂] = 1.50 M, [I₂] = 2.00 M, and [HI] = 3.46 M. What is the equilibrium constant value, Kc, for this reaction at 298 K?

$$H_2(g) + I_2(g) \rightleftharpoons 2\,HI(g)$$

The correct answer is: 4.00.

Units are not used in equilibrium constants. Our interest is Kc's size.	For this reaction $K_c = \dfrac{[HI]^2}{[H_2][I_2]}$ thus $$K_c = \dfrac{[3.46]^2}{[1.50][2.00]} = \dfrac{12.0}{3.00} = 4.00$$	A very common mistake is to forget to properly include the stoichiometric coefficients as exponents.

YIELD

The equilibrium constant value indicates if the reaction favors products, reactants, or both.
1) K_c >10 to 20, the reaction is **product favored**
2) K_c < 1, the reaction is **reactant favored**
3) $1 < K_c < 10$ to 20, the reaction is a **mixture of reactants and products**

INSIGHT: Fundamentally, K_c is a ratio of the product concentrations divided by the reactant concentrations. Which is why the larger the K_c value the more product favored is the reaction. K_c is actually defined using a thermodynamic quantity called activity. Gas activities are the same as their concentrations. In heterogeneous equilibria (involving gases, liquids, and solids) activities of the pure solids and liquids are 1 and can be neglected. Thus K_c for heterogeneous equilibria only require gas concentrations and no solid or liquid contributions.

2. Consider this reaction at 298 K: $H_2(g) + I_2(g) \rightleftharpoons 2\ HI(g)$

The equilibrium constant is 4.00. If the reaction vessel initially has the following reactant concentrations $[H_2]$ = 6.00 M and $[I_2]$ = 4.00 M. What are the equilibrium concentrations of all species in this reaction?

The correct answer is: $[H_2]$ = 3.6 M, $[I_2]$ = 1.6 M, and $[HI]$ = 4.8 M.

INSIGHT:	Problems giving K_c and starting reactant concentrations are solved as follows. Setting up an ICE table is a good approach.

$$H_2(g) + I_2(g) \rightleftharpoons 2\ HI(g)$$

Starting [ ]'s	6.00	4.00	0.00
Change in [ ]	-x	-x	+ 2x
Final [ ]'s	6.00-x	4.00-x	2x

These 2's come from the reaction stoichiometry.

The final [ ]'s will be used in the K_c expression.

$$K_c = \frac{\left[HI\right]^2}{\left[H_2\right]\left[I_2\right]} = 4.00$$

$$= \frac{(2x)^2}{(6.00-x)(4.00-x)} = 4.00$$

Multiplying $(6-x)(4-x)$ gives $24-10x + x^2$.

$$= \frac{4x^2}{24-10x+x^2} = 4.00$$

$$= 4x^2 = 4.00(24-10x+x^2)$$

$$= 4x^2 = 96-40x+4x^2$$

$$= 4x^2 - 4x^2 = 96-40x$$

$$40x = 96$$

$$x = 2.4$$

Use the value of x in the expressions determined above for the final [ ]'s.

$$\text{Final}\left[H_2\right] = 6.00 - x = 6.00 - 2.4 = 3.6M$$

$$\text{Final}\left[I_2\right] = 4.00 - x = 4.00 - 2.4 = 1.6M$$

$$\text{Final}\left[HI\right] = 2x = 2(2.4) = 4.8M$$

INSIGHT:	In this particular problem the x^2 terms cancel out, simplifying the problem. Problems where this does NOT occur are solved using the quadratic equation.

Use of the Equilibrium Constant, K_p

3. *For this reaction at 298 K, the equilibrium partial pressures are:*
$P_{NO_2} = 0.500$ *atm and* $P_{N_2O_4} = 0.0698$ *atm. What is the K_p value for this reaction?*

$$2\ NO_2(g) \rightleftharpoons N_2O_4(g)$$

The correct answer is: $K_p = 0.279$.

$$K_p = \frac{[N_2O_4]}{[NO_2]^2} = \frac{0.0698}{(0.500)^2} = \frac{0.0698}{0.250} = 0.279$$

Stoichiometric coefficients are used in both K_c and K_p calculations.

4. *For this reaction at 298 K, K_p has a value of 0.279, what is the K_c value for this reaction at 298 K?*

$$2\ NO_2(g) \rightleftharpoons N_2O_4(g)$$

The correct answer is: $K_c = 6.84$.

$$K_p = K_c(RT)^{\Delta n} \text{ thus } K_c = \frac{K_p}{(RT)^{\Delta n}}$$

$$\Delta n = \sum \text{moles of gaseous products} - \sum \text{moles of gaseous reactants}$$

$$\Delta n = \text{moles of } N_2O_4 - \text{moles of } NO_2 = 1 - 2 = -1$$

Use the gas law value of R (0.0821 L atm/mol K).

$$K_c = \frac{0.279}{(0.0821 \times 298)^{-1}} = 0.279 \times 24.5 = 6.84$$

$$\frac{1}{(0.0821 \times 298)^{-1}} = \frac{1}{(24.5)^{-1}} = 24.5$$

Effects of Temperature, Pressure, and Concentration on Equilibrium Position

5. *What is the effect of these changes on the equilibrium position for this reaction at 298 K?*

$$2\ NO_2(g) \rightleftharpoons N_2O_4(g)\quad \Delta H^0_{rxn} = -57.2 kJ/mol$$

a) *Increasing the reaction temperature.*
b) *Removing some NO_2 from the reaction vessel.*
c) *Adding some N_2O_4 to the reaction vessel.*
d) *Increasing the pressure in the reaction vessel by adding an inert gas.*
e) *Decreasing by half the reaction vessel.*
f) *Introducing a catalyst into the reaction vessel.*

The correct answers are the position of equilibrium shifts to the: a) left b) left c) left d) no effect e) right f) no effect.

INSIGHT:
> If the equilibrium position shifts to the left, reactant concentrations increase and product concentrations decrease. If the equilibrium position shifts to the right, reactant concentrations decrease and product concentrations increase. All of these changes are illustrations of Le Châtelier's principle: if a system at equilibrium is stressed, it responds to relieve that stress.

a) *For exothermic reactions: increasing the temperature shifts the equilibrium position to the left, decreasing the temperature shifts the equilibrium position to the right. Endothermic reactions behave oppositely.* In exercise 5 the negative ΔH^0_{rxn} indicates an exothermic reaction. A temperature increase shifts the equilibrium position to the left.

b) *If a reactant concentration is decreased below the equilibrium concentration, the equilibrium position changes to restore concentrations back to those predicted by the equilibrium constant.* In exercise 5, removing NO_2 from the reaction vessel decreases $[NO_2]$. The equilibrium responds to the stress by increasing $[NO_2]$ and decreasing $[N_2O_4]$, an equilibrium position shift to the left or reactant side. Adding NO_2 would cause the equilibrium position to shift to the right or product side.

c) *If a product concentration is increased above the equilibrium concentration, the equilibrium position shifts to restore product and reactant concentrations back to those predicted by the equilibrium constant.* Adding some N_2O_4 to the reaction vessel increases $[N_2O_4]$ above the equilibrium concentration. The equilibrium responds to this stress by decreasing $[N_2O_4]$ and increasing $[NO_2]$, an equilibrium position shift to the left or reactant side. Removing N_2O_4 shifts the equilibrium position to the right or product side.

d) *Adding an inert gas to the reaction mixture has no effect on the equilibrium position because the gas concentrations are not affected.* This is a common misconception for students.

e) *If the reaction vessel volume is changed, gas concentrations are changed because for gases, $M \propto n/V$. If the volume is decreased, the equilibrium position shifts to the side having the fewest moles of gas.* In this exercise the right or product side has the fewest moles.

f) *Adding a catalyst has no effect on the equilibrium position.* Catalysts change reaction rates but not equilibrium positions.

Q, The Reaction Quotient

6. *A nonequilibrium mixture has a [NO₂] = 0.50 M and [N₂O₄] = 0.50 M at 298 K. How will the reaction respond to reestablish equilibrium? Will the reactant concentration increase or decrease? Will the product concentration increase or decrease?*

$$2\,NO_2 \rightleftharpoons N_2O_4$$

The correct answer is: product concentration increases as reactant concentration decreases until equilibrium is reestablished.

INSIGHT: | The reaction quotient, Q, is used to predict how nonequilibrium mixtures respond to reestablish equilibrium. Q **uses nonequilibrium concentrations** whereas K_c uses equilibrium concentrations.

$$Q = \frac{[N_2O_4]}{[NO_2]^2} = \frac{[0.50]}{[0.50]^2} = 2.0 \text{ and thus } Q < K_c$$

In exercise 4, we calculated $K_c = 6.84$ for this reaction, thus $Q < K_c$. Because Q is smaller than K_c, the reaction mixture has too few products and too many reactants. (Q is a fraction. Having a value less than K_c indicates the numerator must increase as the denominator decreases to return to K_c.) Thus product concentrations increase and reactant concentrations decrease until equilibrium reestablishes.

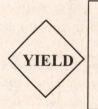

1. If **Q < K_c**, the reaction consumes reactants and yields products to reestablish equilibrium.
2. If **Q > K_c**, the reaction produces reactants and consumes products to reestablish equilibrium.
3. If **Q = K_c**, the reaction is at equilibrium.

Relationship of ΔG^0_{rxn} to K, the Equilibrium Constant

7. *What is the value of the gaseous equilibrium constant, K_p, at 298 K for this reaction?*

$$H_2(g) + F_2(g) \rightleftharpoons 2\,HF(g)$$

The correct answer is: 5.11 x 10⁹⁵.

We calculate ΔG^0_{rxn} using the method described in Module 15. For this reaction ΔG^0_{rxn} = -546 kJ/mol or -5.46 x 10⁵ J/mol.

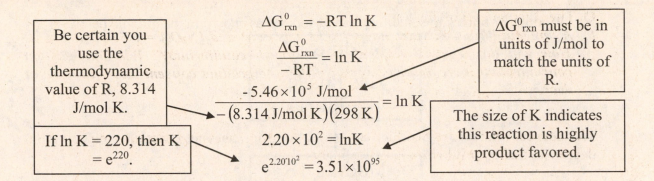

$$\Delta G^0_{rxn} = -RT \ln K$$

$$\frac{\Delta G^0_{rxn}}{-RT} = \ln K$$

$$\frac{-5.46 \times 10^5 \text{ J/mol}}{-(8.314 \text{ J/mol K})(298 \text{ K})} = \ln K$$

$$2.20 \times 10^2 = \ln K$$

$$e^{2.20 \cdot 10^2} = 3.51 \times 10^{95}$$

Be certain you use the thermodynamic value of R, 8.314 J/mol K.

If ln K = 220, then K = e^{220}.

ΔG^0_{rxn} must be in units of J/mol to match the units of R.

The size of K indicates this reaction is highly product favored.

Evaluating Equilibrium Constants at Different Temperatures

8. *This reaction has an equilibrium constant, K_c, of 6.84 at 298 K. What is the K_c value at 225K?*

$$2 NO_2 \rightleftharpoons N_2O_4$$

The correct answer is: 1.15 x 10⁴.

The problem indicates that K_{T_1} = 6.84, T_1 = 298 K, and T_2 = 225 K. From Module 15 we calculate that ΔH^0 = -57.2 kJ/mol = -5.72 x 10⁴ J/mol.

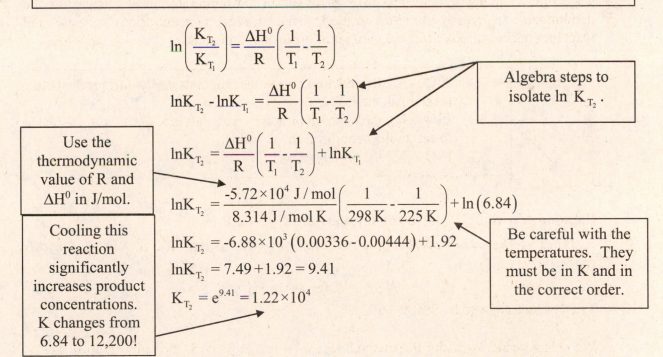

$$\ln\left(\frac{K_{T_2}}{K_{T_1}}\right) = \frac{\Delta H^0}{R}\left(\frac{1}{T_1} - \frac{1}{T_2}\right)$$

$$\ln K_{T_2} - \ln K_{T_1} = \frac{\Delta H^0}{R}\left(\frac{1}{T_1} - \frac{1}{T_2}\right)$$

$$\ln K_{T_2} = \frac{\Delta H^0}{R}\left(\frac{1}{T_1} - \frac{1}{T_2}\right) + \ln K_{T_1}$$

$$\ln K_{T_2} = \frac{-5.72 \times 10^4 \text{ J/mol}}{8.314 \text{ J/mol K}}\left(\frac{1}{298 \text{ K}} - \frac{1}{225 \text{ K}}\right) + \ln(6.84)$$

$$\ln K_{T_2} = -6.88 \times 10^3 (0.00336 - 0.00444) + 1.92$$

$$\ln K_{T_2} = 7.49 + 1.92 = 9.41$$

$$K_{T_2} = e^{9.41} = 1.22 \times 10^4$$

Algebra steps to isolate ln K_{T_2}.

Use the thermodynamic value of R and ΔH^0 in J/mol.

Cooling this reaction significantly increases product concentrations. K changes from 6.84 to 12,200!

Be careful with the temperatures. They must be in K and in the correct order.

Module 17 relates to some following Modules as shown in the graphic below.

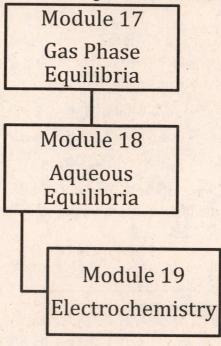

Module 17

Gas Phase Equilibria

Module 18

Aqueous Equilibria

Module 19

Electrochemistry

Module 18 Predictor Questions

The following questions may help you determine the extent you need to study this module. Questions are ranked according to ability.

> Level 1 = basic proficiency
> Level 2 = mid level proficiency
> Level 3 = high proficiency

If you can correctly answer Level 3 questions you probably do not need to spend much time on this module. If you can only answer Level 1 problems, you should review this module.

Level 1	1. What is the hydroxide ion concentration in an aqueous solution with a pH = 6.19?
Level 1	2. What is the pH of an aqueous solution having a hydroxide ion concentration of 5.47×10^{-9}?
Level 1	3. Calculate $[H_3O^+]$ for a 0.010 M HCl solution.
Level 1	4. What are the molar Ca^{2+} and OH^- ion concentrations in a 0.015 M solution of calcium hydroxide? What is the solution pH?
Level 1	5. How many mL of 0.35 M NaOH are required to completely neutralize 20.0 mL of 0.026 M H_2SO_4?
Level 2	6. Calculate the percent ionization and pH of a 0.100 M HNO_2 solution. K_a for HNO_2 is 4.5×10^{-4}.
Level 3	7. What are the pH and pOH of a 0.075 M sodium acetate solution? For acetic acid $K_a = 1.8 \times 10^{-5}$.
Level 1	8. Calculate the pH of a solution that is 0.10 M in acetic acid and 0.30 M in sodium acetate. For acetic acid $K_a = 1.8 \times 10^{-5}$.
Level 3	9. What ratio of NH_4Cl and NH_3 is necessary to make a buffer with pH = 8.50? K_b for NH_3 is 1.8×10^{-5}.
Level 1	10. What is the correct form of the solubility product constant for $Ba_3(AsO_4)_2$?
Level 1	11. If the solubility of BiI_3 is 7.7×10^{-3} g/L and the solubility of $Fe(OH)_2$ is 1.1×10^{-3} g/L (both in water at 25°C), what are the K_{sp} values for each compound?

Module 18 Predictor Question Solutions

1. What is the hydroxide ion concentration in an aqueous solution with a pH = 6.19?

$pH = -log\ [H^+]$
$6.19 = -log\ [H^+]$
$[H^+] = 10^{-6.19} = 6.46 \times 10^{-7}\ M$
$K_w = [OH^-][H^+] = 1.00 \times 10^{-14}\ M$
$[OH^-][6.46 \times 10^{-7}\ M] = 1.00 \times 10^{-14}\ M$
$[OH^-] = 1.55 \times 10^{-8}\ M$

2. What is the pH of an aqueous solution having a hydroxide ion concentration of 5.47 x 10^{-9}?

$pOH = -log[OH^-] = -log[5.47 \times 10^{-9}\ M] = 8.26$
$pOH + pH = 14.00$
$pH = 14.00 - pOH = 14.00 - 8.26$
$pH = 5.74$

3. Calculate $[H_3O^+]$ for a 0.010 M HCl solution.

$$\left(\frac{0.010\ mol\ HCl}{1\ L}\right)\left(\frac{1\ mol\ H^+}{1\ mol\ HCl}\right) = 0.010\ M\ H^+ = 0.010\ M\ H_3O^+$$

4. What are the molar Ca^{2+} and OH^- ion concentrations in a 0.015 M solution of calcium hydroxide? What is the solution pH?

$$\left(\frac{0.015\ mol\ Ca(OH)_2}{L}\right)\left(\frac{1\ mol\ Ca^{2+}}{1\ mol\ Ca(OH)_2}\right) = 0.015\ M\ Ca^{2+}$$

$$\left(\frac{0.015\ mol\ Ca(OH)_2}{L}\right)\left(\frac{2\ mol\ OH^-}{1\ mol\ Ca(OH)_2}\right) = 0.030\ M\ OH^-$$

$pOH = -log[OH^-] = -log[0.030] = 1.52$

$pH = 14.00 - pOH = 14.00 - 1.52 = 12.48$

5. How many mL of 0.35 M NaOH are required to completely neutralize 20.0 mL of 0.026 M H$_2$SO$_4$?

The reaction equation is $H_2SO_4 + 2\ NaOH \rightarrow 2\ H_2O + Na_2SO_4$.

$$20.0\ mL\left(\frac{0.026\ mmol\ H_2SO_4}{mL}\right)\left(\frac{2\ mmol\ NaOH}{1\ mmol\ H_2SO_4}\right)\left(\frac{mL}{0.35\ mmol\ NaOH}\right) = 2.97\ mL$$

6. Calculate the percent ionization and pH of a 0.100 M HNO$_2$ solution. K_a for HNO$_2$ is 4.5×10^{-4}.

$$\text{HNO}_2 \rightleftharpoons \text{H}^+ + \text{NO}_2^-$$

[Initial]	0.100 M	0	0
$\Delta[\ ]$	-x	+x	+x
[Equilibrium]	0.100 M - x	x	x

$$K_a = \frac{\left[\text{H}^+\right]\left[\text{NO}_3^-\right]}{\left[\text{HNO}_2\right]} = \frac{x^2}{0.100 - x}$$

Because x is much smaller than 0.100, the equation

can be simplified to : $K_a = \dfrac{x^2}{0.100} = 4.5 \times 10^{-4}$

$$x = 6.7 \times 10^{-3} = \left[\text{H}^+\right]$$

$$\text{pH} = -\log\left[\text{H}^+\right] = -\log\left(6.7 \times 10^{-3}\right) = 2.17$$

$$\%\ \text{ionization} = \frac{\left[\text{ionized acid}\right]}{\left[\text{initial acid}\right]} \times 100 = \frac{\left[6.7 \times 10^{-3}\right]}{\left[0.100\right]} \times 100 = 6.7\%$$

7. What are the pH and pOH of a 0.075 M sodium acetate solution?
   For acetic acid $K_a = 1.8 \times 10^{-5}$.

$$\text{H}_2\text{O} + \text{CH}_3\text{COO}^- \rightleftharpoons \text{CH}_3\text{COOH} + \text{OH}^-$$

[Initial]	0.075 M	0	0
$\Delta[\ ]$	-x	+x	+x
[Equilibrium]	0.075 M - x	x	x

$$K_w = K_a K_b \text{ thus } K_b = \frac{K_w}{K_a}$$

$$K_b = \frac{1.00 \times 10^{-14}}{1.8 \times 10^{-5}} = 5.6 \times 10^{-10}$$

$$K_b = 5.6 \times 10^{-10} = \frac{x^2}{0.075 - x} \approx \frac{x^2}{0.075}$$

$$x = 6.5 \times 10^{-6} = \left[\text{OH}^-\right]$$

$$\text{pOH} = -\log\left[\text{OH}^-\right] = -\log\left[6.5 \times 10^{-6}\right] = 5.19$$

$$\text{pH} = 14.00 - 5.19 = 8.81$$

8. Calculate the pH of a solution that is 0.10 M in acetic acid and 0.30 M in sodium acetate. For acetic acid $K_a = 1.8 \times 10^{-5}$.

   This is a buffer problem. Use the Henderson-Hasselbalch equation. Since the buffer solution contains a weak acid and its conjugate base, use the pK_a.

$$K_a = 1.8 \times 10^{-5}$$

$$pK_a = -\log(K_a) = 4.74$$

$$pH = pKa + \log \frac{[\text{conjugate base}]}{[\text{acid}]}$$

$$pH = 4.74 + \log \frac{[0.30\ M]}{[0.10\ M]} = 5.22$$

9. What ratio of NH4Cl and NH3 is necessary to make a buffer with pH = 8.50? Kb for NH3 is 1.8 x 10⁻⁵.

This is a buffer problem. Use the Henderson-Hasselbalch equation. Since the buffer solution contains a weak base and its conjugate acid, use the pKb.

$$K_b = 1.8 \times 10^{-5}$$

$$pK_b = -\log \left[1.8 \times 10^{-5} \right] = 4.74$$

$$pOH = 14.00 - pH = 14.00 - 8.50 = 5.50$$

$$pOH = 5.50 = pK_b + \log \frac{\left[\text{conjugate base} \right]}{\left[\text{acid} \right]}$$

$$5.50 = 4.74 + \log \frac{\left[\text{conjugate base} \right]}{\left[\text{acid} \right]}$$

$$0.76 = \log \frac{\left[\text{conjugate base} \right]}{\left[\text{acid} \right]} \quad \text{thus } 10^{0.76} = 5.75 = \frac{\left[\text{conjugate base} \right]}{\left[\text{acid} \right]}$$

10. What is the correct form of the solubility product constant for Ba3(AsO4)2?

$$Ba_3(AsO_4)_2 \rightleftharpoons 3\ Ba^{2+} + 2\ AsO_4^{2-}$$

$$K_{sp} = [Ba^{2+}]^3[AsO_4^{2-}]^2$$

11. If the solubility of BiI3 is 7.7 x 10⁻³ g/L and the solubility of Fe(OH)2 is 1.1 x 10⁻³ g/L (both in water at 25°C), what are the Ksp values for each compound?

$$\left(\frac{7.7 \times 10^{-3}\ g\ BiI_3}{L} \right) \left(\frac{1\ mol\ BiI_3}{589.68\ g\ BiI_3} \right) = 1.31 \times 10^{-5}\ M$$

$$BiI_3 \rightleftharpoons Bi^{3+} + 3\ I^-$$

$$Bi^{3+} = 1.31 \times 10^{-5}\ M$$

$$[I^-] = 3(1.31 \times 10^{-5}\ M) = 3.93 \times 10^{-5}\ M$$

$$K_{sp} = [Bi^{3+}][I^-]^3 = [1.31 \times 10^{-5}][3.93 \times 10^{-5}]^3$$

$$K_{sp} = 7.95 \times 10^{-19}$$

$$\left(\frac{1.1 \times 10^{-3} \text{ g Fe(OH)}_2}{\text{L}}\right)\left(\frac{1 \text{ mol Fe(OH)}_2}{89.87 \text{ g Fe(OH)}_2}\right) = 1.22 \times 10^{-5} \ M$$

$$\text{Fe(OH)}_2 \rightleftharpoons 2 \text{ Fe}^{2+} + 2 \text{ OH}^-$$

$$\text{Fe}^{2+} = 1.22 \times 10^{-5} \ M$$

$$[\text{OH}^-] = 2(1.22 \times 10^{-5} \ M) = 2.44 \times 10^{-5} \ M$$

$$K_{sp} = [\text{Fe}^{3+}][\text{OH}^-]^2 = [1.22 \times 10^{-5}][2.44 \times 10^{-5}]^2$$

$$K_{sp} = 7.26 \times 10^{-15}$$

Module 18
Aqueous Equilibria

Introduction
This module describes calculations required to determine species concentrations in various aqueous solutions. Exercises in this module will include:
1. determining aqueous hydronium and hydroxide ion concentrations
2. calculating pH, pOH, and % ionization of various solutions
3. acid-base titration calculations
4. determining solution pH and pOH in hydrolysis
5. calculating concentrations and pH of buffer solutions
6. calculating ion concentrations for insoluble solids.

Module 18 Key Equations & Concepts

1. The ionization constant for water,
$$K_w = \left[H_3O^+ \right]\left[OH^- \right] = 1.00 \times 10^{-14}$$
$$14 = pH + pOH$$

K_w is used to calculate either the hydronium or hydroxide ion concentration in aqueous solutions given the concentration of either ion. The second equation is mathematically equivalent to the first and relates pH and pOH.

2.
$$pH = -\log\left[H^+ \right]$$
$$pOH = -\log\left[OH^- \right]$$
$$pK_a = -\log K_a$$

In chemistry, the symbol pX is defined as $-\log X$. These three equations define pH, pOH, and pK_a. pH is a condensed method to write H^+ or H_3O^+ concentration in aqueous solutions. pOH is an equivalent method for writing aqueous OH^- concentration. pK_a is a shorthand method for writing K_a values.

For the weak acid equilibrium $HA \rightleftharpoons H^+ + A^-$
$$K_a = \frac{\left[H^+ \right]\left[A^- \right]}{\left[HA \right]}$$

3. For the weak base equilibrium $BOH \rightleftharpoons B^+ + OH^-$
$$K_b = \frac{\left[B^+ \right]\left[OH^- \right]}{BOH}$$

$$\% \text{ ionization} = \frac{\left[\text{ionized species} \right]}{\left[\text{initial species} \right]} \times 100$$

K_a and K_b are ionization constants for weak acids and weak bases, respectively. K_a and K_b are used to determine species concentrations in aqueous weak acid and base solutions. The % ionization is a description of the amount of weak acid and base ionized in solution. K_a and K_b values are tabulated in your textbook appendices.

---

4. Henderson-Hasselbalch equations:

For an acidic buffer solution $pH = pK_a + \log \dfrac{[salt]}{[acid]}$

For a basic buffer solution $pOH = pK_b + \log \dfrac{[salt]}{[base]}$

Both forms of the equation are used to find the pH or pOH of buffer solutions given concentrations of the buffer components.

5. For acid-base conjugate pairs $K_w = K_a \times K_b = 1.00 \times 10^{-14}$

For buffer and hydrolysis calculations, this relationship determines the acid ionization constant for the conjugate acid of a weak base or the base ionization constant for the conjugate base of a weak acid.

6. For a water insoluble solid where $M_y X_z(s) \rightleftharpoons yM^{Z+}(aq) + zX^{Y-}(aq)$

$$K_{sp} = \left[M^{Z+} \right]^y \left[X^{Y-} \right]^z$$

The solubility product constant, K_{sp}, describes the ion concentrations of insoluble solids in aqueous solutions.

Sample Exercises

Water Ionization Constant

1. What is the $[OH^-]$ in an aqueous solution with a pH = 5.25?
  The correct answer is: $[OH^-] = 1.8 \times 10^{-9}$ M.

| pH calculations use logs and exponential powers of 10. | $pH = -\log\left[H^+\right] \Rightarrow \left[H^+\right] = 10^{-pH}$ $\left[H^+\right] = 10^{-5.25} = 5.6 \times 10^{-6}\ M$ $K_w = 1.00 \times 10^{-14} = \left[H^+\right]\left[OH^-\right]$ thus $\dfrac{1.00 \times 10^{-14}}{\left[H^+\right]} = \left[OH^-\right] = \dfrac{1.00 \times 10^{-14}}{5.6 \times 10^{-6}} = 1.8 \times 10^{-9} M$ | Determining $[OH^-]$, given $[H^+]$, from water's ionization constant. |

2. What is the pH of an aqueous solution having $[OH^-] = 3.45 \times 10^{-3}$?
  The correct answer is: pH = 11.538.

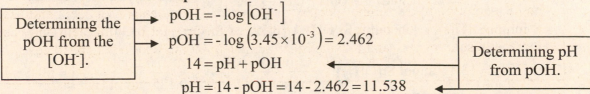

| Determining the pOH from the $[OH^-]$. | $pOH = -\log\left[OH^-\right]$ $pOH = -\log\left(3.45 \times 10^{-3}\right) = 2.462$ $14 = pH + pOH$ $pH = 14 - pOH = 14 - 2.462 = 11.538$ | Determining pH from pOH. |

Strong Acid or Base Dissociation

3. What is the pH of an aqueous 0.025 M Sr(OH)₂ solution?
   The correct answer is: pH = 12.70.

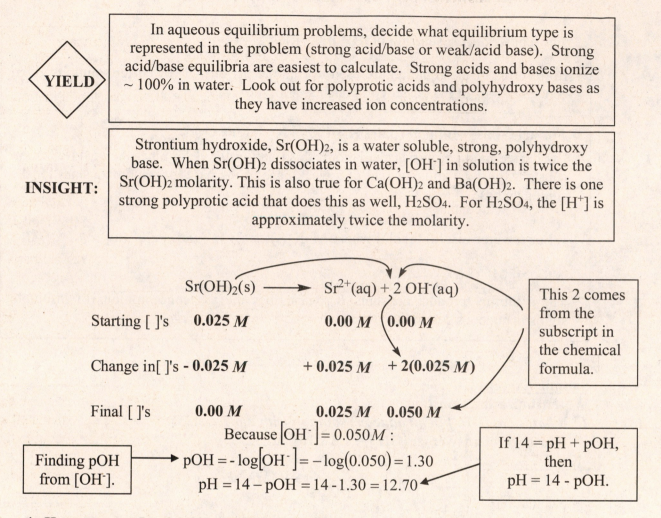

◇ **YIELD** ◇ — In aqueous equilibrium problems, decide what equilibrium type is represented in the problem (strong acid/base or weak/acid base). Strong acid/base equilibria are easiest to calculate. Strong acids and bases ionize ~ 100% in water. Look out for polyprotic acids and polyhydroxy bases as they have increased ion concentrations.

INSIGHT: Strontium hydroxide, Sr(OH)₂, is a water soluble, strong, polyhydroxy base. When Sr(OH)₂ dissociates in water, [OH⁻] in solution is twice the Sr(OH)₂ molarity. This is also true for Ca(OH)₂ and Ba(OH)₂. There is one strong polyprotic acid that does this as well, H₂SO₄. For H₂SO₄, the [H⁺] is approximately twice the molarity.

$$Sr(OH)_2(s) \longrightarrow Sr^{2+}(aq) + 2\ OH^-(aq)$$

Starting [ ]'s   **0.025 M**   **0.00 M**   **0.00 M**

Change in[ ]'s **- 0.025 M**   **+ 0.025 M** **+ 2(0.025 M)**

This 2 comes from the subscript in the chemical formula.

Final [ ]'s   **0.00 M**   **0.025 M**   **0.050 M**

Because $\left[OH^-\right] = 0.050 M$:

Finding pOH from [OH⁻].
$$pOH = -\log\left[OH^-\right] = -\log(0.050) = 1.30$$
$$pH = 14 - pOH = 14 - 1.30 = 12.70$$

If 14 = pH + pOH, then pH = 14 - pOH.

4. How many mL of 0.125 M HCl are required to exactly neutralize 25.0 mL of aqueous 0.025 M Sr(OH)₂ solution?
   The correct answer is: 50.0 mL.

INSIGHT: "Neutralize" is a clue this is a titration problem. Note it is the reaction of a strong acid with a strong dihydroxy base, Sr(OH)₂. In all titrations, the 1ˢᵗ step is to **write a balanced chemical reaction**.

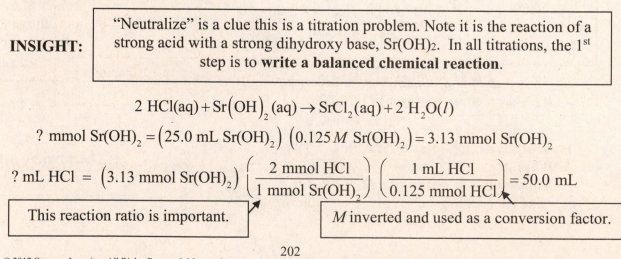

$$2\ HCl(aq) + Sr\left(OH\right)_2 (aq) \rightarrow SrCl_2(aq) + 2\ H_2O(l)$$

$$?\ mmol\ Sr(OH)_2 = \left(25.0\ mL\ Sr(OH)_2\right)\left(0.125\ M\ Sr(OH)_2\right) = 3.13\ mmol\ Sr(OH)_2$$

$$?\ mL\ HCl = \left(3.13\ mmol\ Sr(OH)_2\right)\left(\frac{2\ mmol\ HCl}{1\ mmol\ Sr(OH)_2}\right)\left(\frac{1\ mL\ HCl}{0.125\ mmol\ HCl}\right) = 50.0\ mL$$

This reaction ratio is important.

M inverted and used as a conversion factor.

Weak Acid or Base Ionization

5. What is the pH and % ionization of an aqueous 0.125 M acetic acid, CH₃COOH, solution? For acetic acid $K_a = 1.8 \times 10^{-5}$

   The correct answer is: pH = 2.82 and % ionization =1.2%.

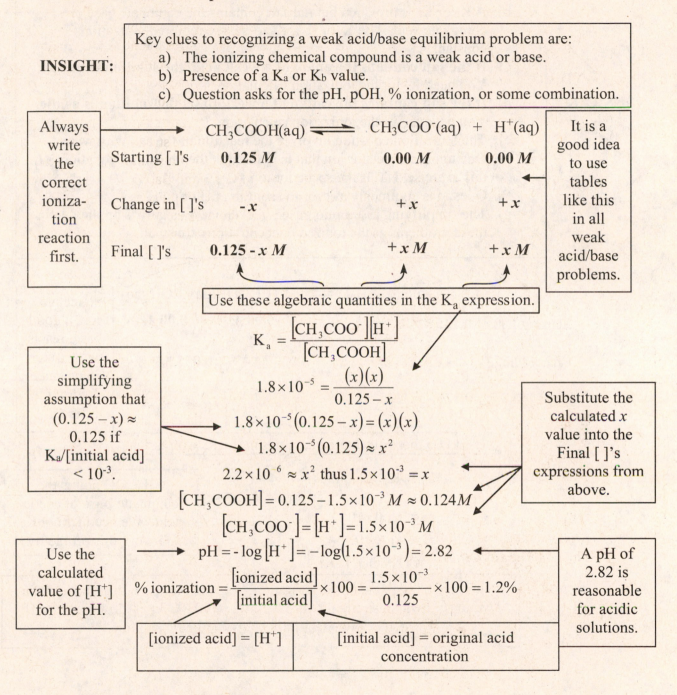

INSIGHT:

Key clues to recognizing a weak acid/base equilibrium problem are:
   a) The ionizing chemical compound is a weak acid or base.
   b) Presence of a K_a or K_b value.
   c) Question asks for the pH, pOH, % ionization, or some combination.

Always write the correct ionization reaction first.

It is a good idea to use tables like this in all weak acid/base problems.

$$CH_3COOH(aq) \rightleftharpoons CH_3COO^-(aq) + H^+(aq)$$

	CH₃COOH	CH₃COO⁻	H⁺
Starting [ ]'s	0.125 M	0.00 M	0.00 M
Change in [ ]'s	-x	+x	+x
Final [ ]'s	0.125 - x M	+x M	+x M

Use these algebraic quantities in the K_a expression.

$$K_a = \frac{[CH_3COO^-][H^+]}{[CH_3COOH]}$$

$$1.8 \times 10^{-5} = \frac{(x)(x)}{0.125 - x}$$

Use the simplifying assumption that $(0.125 - x) \approx 0.125$ if $K_a/[\text{initial acid}] < 10^{-3}$.

$$1.8 \times 10^{-5}(0.125 - x) = (x)(x)$$

$$1.8 \times 10^{-5}(0.125) \approx x^2$$

$$2.2 \times 10^{-6} \approx x^2 \text{ thus } 1.5 \times 10^{-3} = x$$

$$[CH_3COOH] = 0.125 - 1.5 \times 10^{-3} M \approx 0.124 M$$

$$[CH_3COO^-] = [H^+] = 1.5 \times 10^{-3} M$$

Substitute the calculated x value into the Final [ ]'s expressions from above.

Use the calculated value of [H⁺] for the pH.

$$pH = -\log[H^+] = -\log(1.5 \times 10^{-3}) = 2.82$$

$$\% \text{ ionization} = \frac{[\text{ionized acid}]}{[\text{initial acid}]} \times 100 = \frac{1.5 \times 10^{-3}}{0.125} \times 100 = 1.2\%$$

A pH of 2.82 is reasonable for acidic solutions.

[ionized acid] = [H⁺]

[initial acid] = original acid concentration

Hydrolysis or Solvolysis Problem

6. What are the pH and pOH of an aqueous 0.125 M sodium acetate, NaCH₃COO, solution? For acetic acid $K_a = 1.8 \times 10^{-5}$

   The correct answer is: pH = 8.92 and pOH =5.08.

203

Key clues to recognizing a hydrolysis or solvolysis problem:

a) The ionizing chemical compound is a soluble salt of a weak acid or base. ($NaCH_3COO$ is the soluble salt of acetic acid.)

b) A K_a or K_b is provided but the salt contains the conjugate base or conjugate acid. Use $K_w = K_a \times K_b$ to get the required ionization constant.

c) **If the salt contains the anion of a weak acid, the solution is basic.** You need K_b for the ionization calculation.

d) **If the salt contains the cation of a weak base, the solution is acidic.** You need K_a for the ionization calculation.

e) The ionization equation involves the reaction of the salt with water. One ion is a spectator ion that is ignored in the ionization equation. (In this exercise, Na^+ is the spectator ion.)

f) Questions commonly ask for the solution pH or pOH.

g) The simplifying assumption used in exercise 4 usually is applicable in these problems as the ionized concentrations are small.

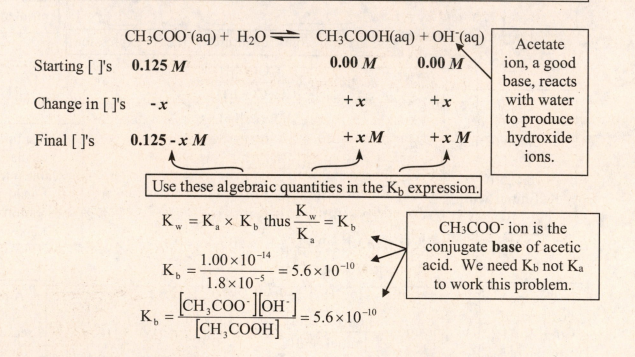

$$CH_3COO^-(aq) + H_2O \rightleftharpoons CH_3COOH(aq) + OH^-(aq)$$

Starting [ ]'s	**0.125 M**	**0.00 M**	**0.00 M**
Change in [ ]'s	**- x**	**+ x**	**+ x**
Final [ ]'s	**0.125 - x M**	**+ x M**	**+ x M**

Acetate ion, a good base, reacts with water to produce hydroxide ions.

Use these algebraic quantities in the K_b expression.

$$K_w = K_a \times K_b \text{ thus } \frac{K_w}{K_a} = K_b$$

$$K_b = \frac{1.00 \times 10^{-14}}{1.8 \times 10^{-5}} = 5.6 \times 10^{-10}$$

$$K_b = \frac{[CH_3COO^-][OH^-]}{[CH_3COOH]} = 5.6 \times 10^{-10}$$

CH_3COO^- ion is the conjugate **base** of acetic acid. We need K_b not K_a to work this problem.

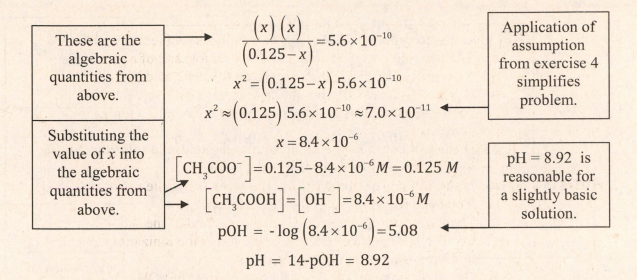

These are the algebraic quantities from above.

$$\frac{(x)(x)}{(0.125-x)}=5.6\times10^{-10}$$

$$x^2=(0.125-x)\,5.6\times10^{-10}$$

$$x^2\approx(0.125)\,5.6\times10^{-10}\approx7.0\times10^{-11}$$

Application of assumption from exercise 4 simplifies problem.

$$x=8.4\times10^{-6}$$

Substituting the value of x into the algebraic quantities from above.

$$[CH_3COO^-]=0.125-8.4\times10^{-6}\,M=0.125\,M$$

$$[CH_3COOH]=[OH^-]=8.4\times10^{-6}\,M$$

pH = 8.92 is reasonable for a slightly basic solution.

$$pOH=-\log(8.4\times10^{-6})=5.08$$

$$pH=14-pOH=8.92$$

Buffer Solution Problem

7. *What are the concentrations of the relevant species and pH of a solution that is 0.100 M in acetic acid, CH₃COOH, and 0.025 M in sodium acetate, NaCH₃COO? For acetic acid $K_a = 1.8 \times 10^{-5}$*

   The correct answer is: [CH₃COOH] = 0.100 M, [CH₃COO⁻] = 0.025 M, [H⁺] = 7.2 x 10⁻⁵ M, and pH = 4.14.

INSIGHT:

Key clues to recognizing buffer problems:
a) Solution contains a soluble salt dissolved in either a weak acid or weak base. Concentrations or amounts of both the salt and weak acid or base are in the problem.
b) Salt is the conjugate partner of the weak acid or base.
c) Equilibrium table has starting concentrations of both salt and acid or base. Equilibrium involves the common ion effect.
d) Henderson-Hasselbalch equations are a simple method to find the buffer solution's pH.
e) Simplifying assumption is useful to quickly determine the solution concentration problem.

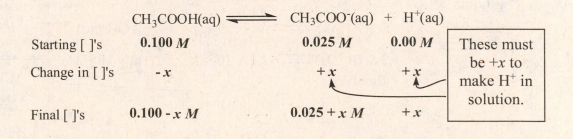

	CH₃COOH(aq) ⇌	CH₃COO⁻(aq) +	H⁺(aq)	
Starting [ ]'s	0.100 M	0.025 M	0.00 M	These must be +x to make H⁺ in solution.
Change in [ ]'s	-x	+x	+x	
Final [ ]'s	0.100 - x M	0.025 + x M	+x	

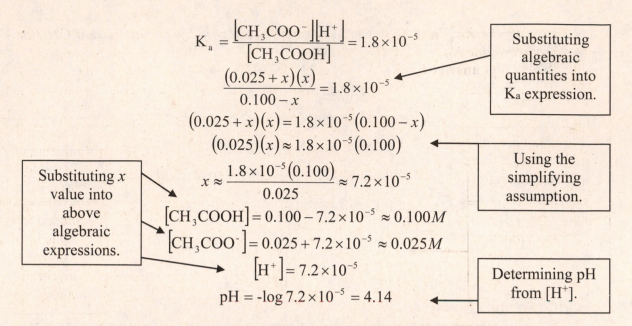

$$K_a = \frac{[CH_3COO^-][H^+]}{[CH_3COOH]} = 1.8 \times 10^{-5}$$

Substituting algebraic quantities into K_a expression.

$$\frac{(0.025 + x)(x)}{0.100 - x} = 1.8 \times 10^{-5}$$

$$(0.025 + x)(x) = 1.8 \times 10^{-5}(0.100 - x)$$

$$(0.025)(x) \approx 1.8 \times 10^{-5}(0.100)$$

Using the simplifying assumption.

Substituting x value into above algebraic expressions.

$$x \approx \frac{1.8 \times 10^{-5}(0.100)}{0.025} \approx 7.2 \times 10^{-5}$$

$$[CH_3COOH] = 0.100 - 7.2 \times 10^{-5} \approx 0.100\,M$$

$$[CH_3COO^-] = 0.025 + 7.2 \times 10^{-5} \approx 0.025\,M$$

$$[H^+] = 7.2 \times 10^{-5}$$

$$pH = -\log 7.2 \times 10^{-5} = 4.14$$

Determining pH from $[H^+]$.

If the question only asks for the buffer solution pH, the simplest method is to use the Henderson-Hasselbalch equations.

Solubility Product Problem

8. The aqueous solubility of iron(II) hydroxide, Fe(OH)$_2$ is 1.1 x 10^{-3} g/L at 25.0°C. What is the solubility product constant for Fe(OH)$_2$?
The correct answer is: K$_{sp}$ = 6.9 x 10^{-15}.

INSIGHT:
Key clues to recognizing solubility product problems:
a) Problem has an insoluble salt dissolved in water.
b) Usually a K_{sp} value is given or sought.
c) In calculating K_{sp} value remember to include stoichiometric factors in both the equilibrium and K_{sp} calculations.
d) Ion concentrations and K_{sp} values are quite small.

Converting g/L into M for K_{sp}.

$$M = \left(\frac{1.1 \times 10^{-3}\ g}{L}\right)\left(\frac{1\ mol\ Fe(OH)_2}{89.87\ g}\right) = 1.2 \times 10^{-5}\,M$$

Set up the dissociation reaction properly.

$$Fe(OH)_2(s) \rightleftharpoons Fe^{2+}(aq) + 2\ OH^-(aq)$$

1.2 x 10^{-5} M $\Longrightarrow$ 1.2 x 10^{-5} M 2(1.2 x 10^{-5} M)
(dissolved)

Do not forget stoichiometric coefficients in both places.

$$K_{sp} = [Fe^{2+}][OH^-]^2$$

$$K_{sp} = (1.2 \times 10^{-5})(2.4 \times 10^{-5})^2$$

$$K_{sp} = 6.9 \times 10^{-15}$$

9. What are the Zn^{2+} and OH^- molar solubilities for zinc hydroxide at 25.0°C? For zinc hydroxide $K_{sp} = 4.5 \times 10^{-17}$

The correct answer is: $[Zn^{2+}] = 2.2 \times 10^{-5}$ M, $[OH^-] = 4.4 \times 10^{-5}$ M.

Dissociation reaction indicates proper concentrations.		Substitute algebraic expressions into K_{sp}.

$$Zn(OH)_2(s) \rightleftharpoons Zn^{2+}(aq) + 2\,OH^-(aq)$$
$$x\,M \Longrightarrow +x\,M \quad + 2x\,M$$

$$K_{sp} = [Zn^{2+}][OH^-]^2 = 4.5 \times 10^{-17}$$
$$K_{sp} = (x)(2x)^2 = 4x^3 = 4.5 \times 10^{-17}$$

K_{sp} calculations often require cube roots or higher.		Substitute x value into above algebraic expressions.

$$x^3 = \frac{4.5 \times 10^{-17}}{4} = 1.1 \times 10^{-17}$$
$$x = \sqrt[3]{1.1 \times 10^{-17}} = 2.2 \times 10^{-6}$$
$$[Zn^{2+}] = x = 2.2 \times 10^{-6}\,M$$
$$[OH^-] = 2x = 4.4 \times 10^{-6}\,M$$

Tips on aqueous equilibrium problems:

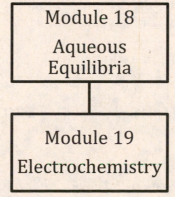

a) The hardest part is deciding if the problem is a weak acid/base, a solvolysis, a buffer, or a solubility product problem. **INSIGHT** columns provide key clues to the problem type encountered. Be very familiar with these.

b) Write the correct ionization reaction and ionization expression (K_a, K_b, K_{sp}, etc.) for the given problem then set up the appropriate ICE table underneath the ionization reaction to help with algebra.

c) Correct use of simplifying assumption saves enormous amounts of time and yields the correct answer. Use it whenever you can!

d) Every aqueous equilibrium problem uses the same basic mathematical method. Major differences are setting up ionization reactions and placing concentrations in appropriate places in table. Once you recognize the problem type solving it is straightforward.

e) The hardest equilibrium for students to recognize is always hydrolysis problems. Keep an eye out for these.

Module 18 relates to the following Module as shown in the graphic below.

Module 18

Aqueous Equilibria

Module 19

Electrochemistry

Practice Test Five
Modules 16-18

Level 1   1. Using the experimental data provided, determine the rate-law expression for this reaction.

$$2 A + B + C \rightarrow D + E$$

Trial	Initial [A]	Initial [B]	Initial [C]	Initial rate of formation of BC
1	0.20 M	0.20 M	0.20 M	$2.4 \times 10^{-6} \ M \cdot min^{-1}$
2	0.40 M	0.30 M	0.20 M	$9.6 \times 10^{-6} \ M \cdot min^{-1}$
3	0.20 M	0.30 M	0.20 M	$2.4 \times 10^{-6} \ M \cdot min^{-1}$
4	0.20 M	0.40 M	0.60 M	$7.2 \times 10^{-6} \ M \cdot min^{-1}$

Level 1   2. The reaction $2 N_2O_5(g) \rightarrow 2N_2O_4(g) + O_2(g)$ has a rate constant of $0.00840 \ s^{-1}$. If 2.25 mol of N_2O_5 are placed in a 4.00 L container, what is the concentration of N_2O_5 after 2.50 minutes? Is the reaction first or second order with respect to N_2O_5?

Level 1   3. Decomposition of NOBr(g) to NO(g) and $Br_2(g)$ has $k = 0.810 \ M^{-1}s^{-1}$. If the initial NOBr concentration is $4.00 \times 10^{-3} \ M$, how long will it take for the NOBr concentration to decrease to $1.50 \times 10^{-4} \ M$?

Level 1   4. Calculate E_a for a reaction having a rate constant of $1.2 \times 10^2 \ s^{-1}$ at 273 K and a second rate constant of $3.6 \times 10^2 \ s^{-1}$ at 298 K.

Level 1   5. Calculate K_c for the following reaction given the equilibrium concentrations of $[F_2] = 1.8 \times 10^{-3} \ M$; $[Br_2] = 9.0 \times 10^{-3} \ M$; $[BrF_5] = 4.6 \times 10^{-3} \ M$.

$$2 BrF_5(g) \rightleftharpoons Br_2(g) + 5 F_2(g)$$

Level 3   6. At a given temperature, K_c for this reaction is 0.0104. If the initial PCl_5 concentration is 0.55 M what is the equilibrium PCl_3 concentration?

$$PCl_5(g) \rightleftharpoons PCl_3(g) + Cl_2(g)$$

Level 2   7. If a reaction is endothermic, will increasing the reaction temperature favor products or reactants? Explain.

Level 1   8. What is the pH of a solution with a hydroxide ion concentration of 1.83×10^{-7}?

Level 3   9. Determine the pH and the pOH of a 0.025 M potassium nitrite solution. K_a for HNO_2 is 4.5×10^{-4}.

Level 3   10. What is the pH of a solution having a ratio of benzoic acid to sodium benzoate of 2:1. K_a for benzoic acid is 6.3×10^{-5}.

Module 19 Predictor Questions

The following questions may help you determine the extent you need to study this module. Questions are ranked according to ability.

Level 1 = basic proficiency
Level 2 = mid level proficiency
Level 3 = high proficiency

If you can correctly answer Level 3 questions you probably do not need to spend much time on this module. If you can only answer Level 1 problems, you should review this module.

Level 3   1. Balance the following oxidation-reduction reaction in acidic solution. Identify the oxidized and reduced species as well as the oxidizing and reducing agents.

$$I_2(s) + S_2O_3^{2-}(aq) \rightarrow I^-(s) + S_4O_6^{2-}(aq)$$

Level 3   2. Balance the following oxidation-reduction reaction in basic solution. Identify the oxidized and reduced species as well as the oxidizing and reducing agents.

$$CrO_2^-(aq) + ClO^-(aq) \rightarrow CrO_4^{2-}(aq) + Cl^-(aq)$$

Level 2   3. What mass of iron metal is produced at the cathode after 2.50 amps of current pass through an electrolytic cell containing iron (II) nitrate for 65 minutes?

Level 2   4. A voltaic cell contains a strip of Ni metal immersed in 1.0 M Ni(NO$_3$)$_2$ solution and a strip of Cu metal immersed in 1.0 M CuSO$_4$ solution. Draw the cell and determine what chemical species are produced at the cathode and anode. What is the direction of electron flow in the cell?

Level 1   5. Determine the standard cell potential for the voltaic cell in question 4.

Level 2   6. What is the cell potential at 338 K for the voltaic cell in question 4 if the Ni(NO$_3$)$_2$ concentration is 2.5 M and the CuSO$_4$ concentration is 1.3 M?

Level 1   7. Determine the equilibrium constant for the voltaic cell in question 4 at 298 K.

Module 19 Predictor Question Solutions

1. Balance the following oxidation-reduction reaction in acidic solution. Identify the oxidized and reduced species as well as the oxidizing and reducing agents.

$$I_2(s) + S_2O_3^{2-}(aq) \rightarrow I^-(s) + S_4O_6^{2-}(aq)$$

$I_2 + 2e^- \rightarrow 2I^-$	**reduction half-reaction**
$2\ S_2O_3^{2-} \rightarrow S_4O_6^{2-} + 2e^-$	**oxidation half-reaction**
$I_2 + 2S_2O_3^{2-} \rightarrow 2I^- + S_4O_6^{2-}$	**complete balanced reaction**

I_2 is reduced and the oxidizing agent.
$S_2O_3^{2-}$ is oxidized and the reducing agent.

2. Balance the following oxidation-reduction reaction in basic solution. Identify the oxidized and reduced species as well as the oxidizing and reducing agents.

$$CrO_2^-(aq) + ClO^-(aq) \rightarrow CrO_4^{2-}(aq) + Cl^-(aq)$$

$(ClO^- + 2\ H^+ + 2e^- \rightarrow Cl^- + H_2O)3$	**reduction half-reaction**
$(CrO_2^- + 2\ H_2O \rightarrow CrO_4^{2-} + 4\ H^+ + 3e^-)2$	**oxidation half-reaction**

$2\ CrO_2^- + 4\ H_2O + 3\ ClO^- + 6\ H^+ \rightarrow 2\ CrO_4^{2-} + 8\ H^+ + 3Cl^- + 3H_2O$ which reduces to $2\ CrO_2^- + H_2O + 3\ ClO^- \rightarrow 2\ CrO_4^{2-} + 2\ H^+ + 3Cl^-$, the balanced reaction in acidic solution.
Since the reaction takes place in basic solution, H^+ cannot exist in large concentrations. Instead H^+ combines with OH^- to form water. To balance in basic solution add two OH^- to both reaction sides giving this balanced equation:

$$2\ CrO_2^- + 3ClO^- + 2\ OH^- \rightarrow 2\ CrO_4^{2-} + 3Cl^- + H_2O$$

3. What mass of iron metal is produced at the cathode after 2.50 amps of current pass through an electrolytic cell containing iron (II) nitrate for 65 minutes?

$$2.50 \text{ amps} = \frac{2.50 \text{ C}}{\text{s}}$$

$$(65 \text{ min})\left(\frac{60 \text{ s}}{1 \text{ min}}\right) = (3900 \text{ s})$$

$$(3900 \text{ s})\left(\frac{2.50 \text{ C}}{\text{s}}\right) = 9750 \text{ C}$$

$$(9750 \text{ C})\left(\frac{1 \text{ mol e}^-}{9.65 \text{ x } 10^4 \text{ C}}\right)\left(\frac{1 \text{ mol Fe}}{2 \text{ mol e}^-}\right)\left(\frac{55.85 \text{ g Fe}}{1 \text{ mol Fe}}\right) = 2.82 \text{ g Fe(s)}$$

4. A voltaic cell contains a strip of Ni metal immersed in 1.0 M Ni(NO$_3$)$_2$ solution and a strip of Cu metal immersed in 1.0 M CuSO$_4$ solution. Draw the cell and determine what chemical species are produced at the cathode and anode. What is the direction of electron flow in the cell?

In voltaic cells, the oxidation reaction has a more negative standard reduction potential and the reduction reaction has a more positive standard reduction potential.

$$Ni^{2+} + 2e^- \rightarrow Ni_{(s)} \qquad E^0 = -0.25V \qquad \text{more negative = oxidation}$$
$$Cu^{2+} + 2e^- \rightarrow Cu_{(s)} \qquad E^0 = 0.337V \qquad \text{more positive = reduction}$$

$Ni_{(s)}$ is oxidized to Ni^{2+} at the anode, and Cu^{2+} is reduced to $Cu_{(s)}$ at the cathode. Electron flow is from the anode to the cathode (as in all voltaic cells).

5. Determine the standard cell potential for the voltaic cell in question 4.
   The sign of E^0 for the oxidation half-reaction must be reversed, along with the reaction.

$$Ni_{(s)} \rightarrow Ni^{2+} + 2e^- \qquad\qquad E^0 = +0.250V$$
$$\underline{Cu^{2+} + 2e^- \rightarrow Cu_{(s)} \qquad\qquad E^0 = +0.337V}$$
$$Ni_{(s)} + Cu^{2+} \rightarrow Ni^{2+} + Cu_{(s)} \qquad E^0_{cell} = 0.587V$$

6. What is the cell potential at 338 K for the voltaic cell in question 4 if the $Ni(NO_3)_2$ concentration is 2.5 M and the $CuSO_4$ concentration is 1.3 M.

$$E = E^0 - \left(\frac{2.303RT}{nF} \right)\left(\log \frac{[\text{oxidized species}]}{[\text{reduced species}]} \right)$$

$$E = 0.587V - \left(\frac{(2.303)(8.314\frac{J}{mol \cdot K})(338\ K)}{(2\ mol\ e^-)(9.65 \times 10^4 \frac{J}{V \cdot mol\ e^-})} \right)\left(\log\frac{[2.5]}{[1.3]} \right) = 0.577\ V$$

7. Determine the equilibrium constant for the voltaic cell in question 4 at 298 K.

$$nFE^0_{cell} = RT \ln K \Rightarrow \ln K = \frac{nFE^0_{cell}}{RT}$$

$$\ln K = \frac{(2\ mol\ e^-)(9.65 \times 10^4 \frac{J}{V \cdot mol\ e^-})(0.587\ V)}{(8.314\frac{J}{mol \cdot K})(298\ K)} = 45.73$$

$$K = 7.22 \times 10^{19}$$

Module 19
Electrochemistry

Introduction

This module describes some basic electrochemistry methods used in typical textbooks. The module goals are to describe how to:

1. balance redox reactions in acidic and basic solutions
2. discern between electrolytic and voltaic cells
3. use Faraday's law to determine the amount of species reduced in an electrolytic cell
4. determine the anode, cathode, and electron flow in both cell types
5. determine the standard cell potential for a voltaic cell
6. use the Nernst equation to find the cell potential at nonstandard conditions
7. determine the Gibb's Free Energy change and equilibrium constant for a cell from its standard potential.

You will need your textbook opened to the appendix containing the standard electrode potentials to understand this material.

Module 19 Key Equations & Concepts

1. **Electrolytic cells are electrochemical cells where nonspontaneous chemical reactions are forced to occur through application of an external voltage.**

   In electrolytic cells reactions that would not occur in nature, such as the electrolysis of chemical compounds or the electroplating of metals, are forced to occur by the passage of electricity.

2. **Voltaic cells are electrolytic cells where spontaneous chemical reactions occur and electrons generated in the reaction are passed through an external wire.**

   Voltaic cells are batteries such as dry cells and lithium batteries used in watches, cameras, etc.

3. **The anode is the electrode where oxidation occurs in both electrolytic and voltaic cells. The cathode is the electrode where reduction occurs in both cell types.**

   In electrolytic cells the anode is the positive electrode and the cathode is the negative electrode. This is reversed for voltaic cells where the anode is the negative electrode and the cathode is the positive electrode.

4. **Faraday's law of electrolysis describes that the amount of a chemical compound oxidized or reduced at an electrode during electrolysis is directly proportional to the amount of electricity passed through the cell.**

   This law is used to calculate the number of grams of chemical compound transformed from oxidized to reduced species, or vice versa, in an electrolytic cell. Faraday's constant, 1 faraday = 9.65×10^4 coulombs, is essential in Faraday's law calculations.

---

5. **Standard Cell Potentials are the initial voltage produced in a voltaic cell at standard conditions.**

To calculate the standard cell potential, add the standard cell potential for the reduction step to the reverse of the standard cell potential for the oxidation step. Standard cell potentials are tabulated in your textbook appendices.

6. $$E = E^0 - \frac{2.303\ RT}{n\ F} \log Q \text{ or } E = E^0 - \frac{0.0592}{n} \log Q \text{ at } 25.0^\circ C$$

The Nernst equation is used to calculate cell potentials at nonstandard conditions. E is the nonstandard cell potential, E^0 is the standard cell potential, n is the number of moles of electrons in the reaction, F is Faraday's constant, and Q is the reaction quotient.

7. $$\Delta G^0 = -nFE^0_{cell} \text{ or } nFE^0_{cell} = RT \ln K$$

This equation is used to determine either the Gibbs Free Energy change or the equilibrium constant of a chemical reaction from the cell potential.

<u>**Sample Exercises**</u>
Balancing Redox Reactions in Acidic and Basic Solutions
1. Balance the following redox reaction in acidic solution.

$$Cu(s) + NO_3^-(aq) \rightarrow Cu^{2+}(aq) + NO_2(g)$$

The correct answer is:

$$Cu(s) + 4\ H^+(aq) + 2\ NO_3^-(aq) \rightarrow Cu^{2+}(aq) + 2\ NO_2(g) + 2\ H_2O(l).$$

INSIGHT: Redox reactions occur in either acidic or basic solutions. In acidic solutions add H^+ and H_2O to balance the reaction. In basic solutions add OH^- and H_2O. The problem will either state the reaction is in acidic solution, or H^+ will be present as a reactant or product. Similarly, for basic solutions look for a statement that the reaction is in basic solution or the presence of OH^-.

 YIELD

There are two simple methods to balance redox reactions, the change in oxidation number method and the half-reaction method. The change in oxidation number method is more physically correct but the half-reaction method is simpler to learn and more straight forward. Use the half-reaction method. Review the rules for assigning oxidation numbers and balancing redox reactions in your textbook.

213

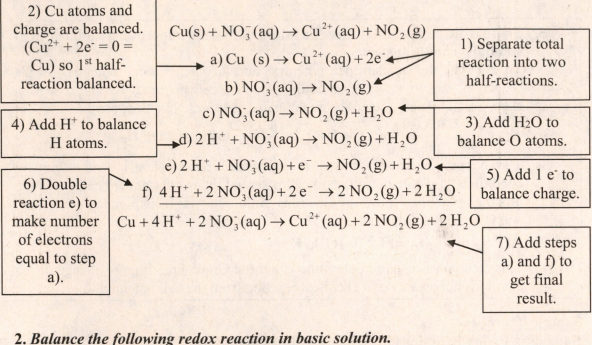

2) Cu atoms and charge are balanced. ($Cu^{2+} + 2e^- = 0 = Cu$) so 1st half-reaction balanced.

1) Separate total reaction into two half-reactions.

$$Cu(s) + NO_3^-(aq) \rightarrow Cu^{2+}(aq) + NO_2(g)$$

a) $Cu\ (s) \rightarrow Cu^{2+}(aq) + 2e^-$

b) $NO_3^-(aq) \rightarrow NO_2(g)$

c) $NO_3^-(aq) \rightarrow NO_2(g) + H_2O$

3) Add H_2O to balance O atoms.

4) Add H^+ to balance H atoms.

d) $2H^+ + NO_3^-(aq) \rightarrow NO_2(g) + H_2O$

e) $2H^+ + NO_3^-(aq) + e^- \rightarrow NO_2(g) + H_2O$

5) Add 1 e^- to balance charge.

6) Double reaction e) to make number of electrons equal to step a).

f) $4H^+ + 2NO_3^-(aq) + 2e^- \rightarrow 2NO_2(g) + 2H_2O$

$$Cu + 4H^+ + 2NO_3^-(aq) \rightarrow Cu^{2+}(aq) + 2NO_2(g) + 2H_2O$$

7) Add steps a) and f) to get final result.

2. Balance the following redox reaction in basic solution.

$$CrO_2^-(aq) + ClO^-(aq) \rightarrow CrO_4^-(aq) + Cl^-(aq)$$

The correct answer is: $CrO_2^-(aq) + 2ClO^-(aq) \rightarrow CrO_4^-(aq) + 2Cl^-(aq)$.

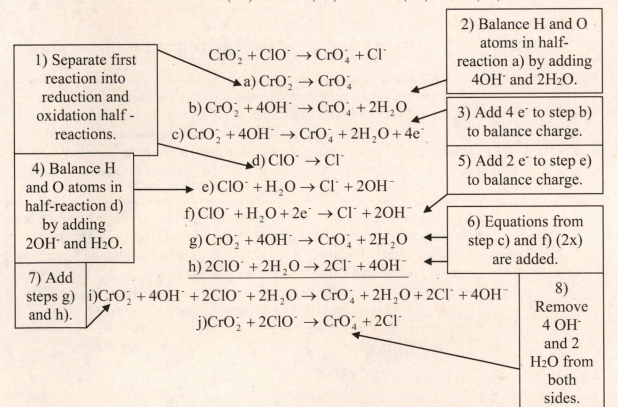

1) Separate first reaction into reduction and oxidation half-reactions.

2) Balance H and O atoms in half-reaction a) by adding $4OH^-$ and $2H_2O$.

$$CrO_2^- + ClO^- \rightarrow CrO_4^- + Cl^-$$

a) $CrO_2^- \rightarrow CrO_4^-$

b) $CrO_2^- + 4OH^- \rightarrow CrO_4^- + 2H_2O$

c) $CrO_2^- + 4OH^- \rightarrow CrO_4^- + 2H_2O + 4e^-$

3) Add 4 e^- to step b) to balance charge.

d) $ClO^- \rightarrow Cl^-$

4) Balance H and O atoms in half-reaction d) by adding $2OH^-$ and H_2O.

e) $ClO^- + H_2O \rightarrow Cl^- + 2OH^-$

5) Add 2 e^- to step e) to balance charge.

f) $ClO^- + H_2O + 2e^- \rightarrow Cl^- + 2OH^-$

g) $CrO_2^- + 4OH^- \rightarrow CrO_4^- + 2H_2O$

6) Equations from step c) and f) (2x) are added.

h) $2ClO^- + 2H_2O \rightarrow 2Cl^- + 4OH^-$

7) Add steps g) and h).

i) $CrO_2^- + 4OH^- + 2ClO^- + 2H_2O \rightarrow CrO_4^- + 2H_2O + 2Cl^- + 4OH^-$

8) Remove 4 OH^- and 2 H_2O from both sides.

j) $CrO_2^- + 2ClO^- \rightarrow CrO_4^- + 2Cl^-$

214

Electrolytic Cell Problem

3. An electrolytic cell containing an aqueous NaCl solution is constructed. What chemical species are produced at the cathode and anode? What is the electron flow direction in this cell?

The correct answer is: $H_2(g)$ is produced at the cathode and $Cl_2(g)$ is produced at the anode. Electrons flow from the anode passing through the battery to the cathode.

INSIGHT:

In this electrolytic cell there are four possible redox reactions. The two possible reductions are $Na^+(aq)$ to $Na(s)$ or H_2O to $H_2(g)$ and $OH^-(aq)$. The two possible oxidations are $Cl^-(aq)$ to $Cl_2(g)$ or H_2O to $H^+(aq)$ and O_2.

How can you determine the correct electrolytic cell reactions? Electrolytic cells force nonspontaneous chemical reactions to occur. Standard reduction potentials from your textbook guide our choices. 1) The correct oxidation reaction has the most positive reduction (most negative oxidation) potential of the possible oxidation reactions. 2) The reduction reaction has the most positive reduction (least negative oxidation) potential of the possible reduction reactions.

In **electrolytic cells**, the - electrode is the cathode (reduction), + electrode is the anode (oxidation).

Reduction Potentials
$$2H_2O + 2e^- \rightarrow H_2 + 4H^+ + 4e^-$$
$$E^0_{cell} = -0.828 \text{ V}$$
$$Na^+ + e^- \rightarrow Na$$
$$E^0_{cell} = -2.71 \text{ V}$$
-0.828 is more positive than -2.71. The 1st reaction occurs.

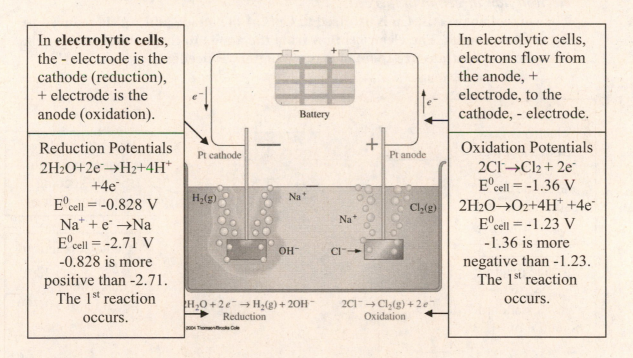

In electrolytic cells, electrons flow from the anode, + electrode, to the cathode, - electrode.

Oxidation Potentials
$$2Cl^- \rightarrow Cl_2 + 2e^-$$
$$E^0_{cell} = -1.36 \text{ V}$$
$$2H_2O \rightarrow O_2 + 4H^+ + 4e^-$$
$$E^0_{cell} = -1.23 \text{ V}$$
-1.36 is more negative than -1.23. The 1st reaction occurs.

Faraday's Law

4. How many grams of nickel metal are produced at the cathode when 3.75 amps of current are passed for 75.0 minutes through an electrolytic cell containing $NiSO_{4(aq)}$?

The correct answer is: 5.14 g of Ni.

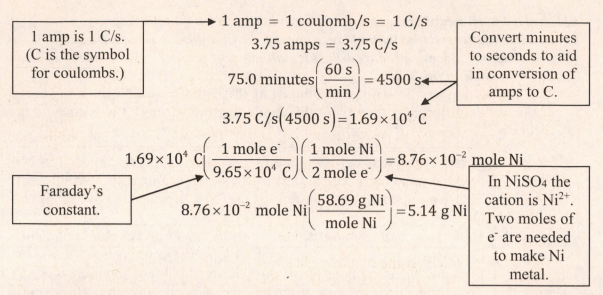

1 amp is 1 C/s. (C is the symbol for coulombs.)

$$1 \text{ amp} = 1 \text{ coulomb/s} = 1 \text{ C/s}$$

$$3.75 \text{ amps} = 3.75 \text{ C/s}$$

$$75.0 \text{ minutes}\left(\frac{60 \text{ s}}{\text{min}}\right) = 4500 \text{ s}$$

Convert minutes to seconds to aid in conversion of amps to C.

$$3.75 \text{ C/s}\left(4500 \text{ s}\right) = 1.69 \times 10^4 \text{ C}$$

$$1.69 \times 10^4 \text{ C}\left(\frac{1 \text{ mole e}^-}{9.65 \times 10^4 \text{ C}}\right)\left(\frac{1 \text{ mole Ni}}{2 \text{ mole e}^-}\right) = 8.76 \times 10^{-2} \text{ mole Ni}$$

Faraday's constant.

$$8.76 \times 10^{-2} \text{ mole Ni}\left(\frac{58.69 \text{ g Ni}}{\text{mole Ni}}\right) = 5.14 \text{ g Ni}$$

In $NiSO_4$ the cation is Ni^{2+}. Two moles of e$^-$ are needed to make Ni metal.

Voltaic Cell

5. A voltaic cell is constructed of a 1.0 M CuSO₄(aq) solution and a 1.0 M AgNO₃(aq) solution with the appropriate metal electrodes, connecting wires, and salt bridges. What chemical species is produced at the cathode and anode? What is the electron flow direction in this cell?

The correct answer is: Cu is oxidized to Cu^{2+} at the anode and Ag^+ is reduced to Ag at the cathode. The electrons flow from the anode to the cathode in this cell. Because the reactions are spontaneous, no battery is needed.

In voltaic cells electrons flow from the anode (negative electrode) to the cathode (positive electrode).

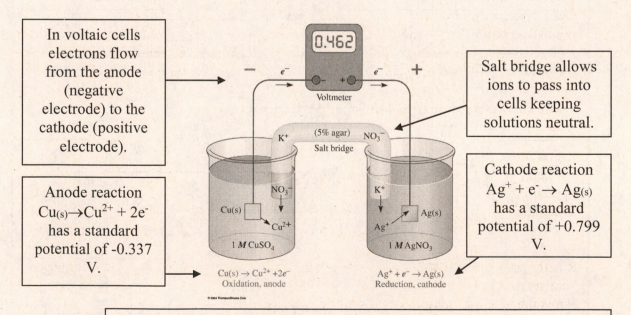

Salt bridge allows ions to pass into cells keeping solutions neutral.

Anode reaction $Cu_{(s)} \rightarrow Cu^{2+} + 2e^-$ has a standard potential of -0.337 V.

Cathode reaction $Ag^+ + e^- \rightarrow Ag_{(s)}$ has a standard potential of +0.799 V.

INSIGHT: In voltaic cells reactions are spontaneous. Predict the reactions using standard reduction potentials. 1) The oxidation reaction has the least positive (most negative) standard reduction potential. 2) The reduction reaction has the most positive (least negative) standard reduction potential.

216

6. What is the standard cell potential for the voltaic cell in Sample Exercise 5?
   The correct answer is: 0.462 V.

<table>
<tr>
<td>
The reduction reaction is balanced **but the E^0_{cell} is not doubled**. Cell potentials are intensive quantities.
</td>
<td>

$2\,Ag^+(aq)+2\,e^- \rightarrow 2\,Ag(s) \qquad\qquad E^0 = +0.799\ V$

$Cu(s) \rightarrow Cu^{2+}(aq)+2\,e^- \qquad\qquad E^0 = -0.337\ V$

$\overline{2\,Ag^+(aq)+Cu(s) \rightarrow 2\,Ag_{(s)} + Cu^{2+}(aq)\ \ E^0_{cell} = +0.462\ V}$

Cell potentials, E^0, are from textbook appendix tables. For voltaic cells, the overall cell potential E^0_{cell} must be a **positive** value.
</td>
<td>
Reverse the reaction and change the sign of the tabulated reduction potential for the oxidation reaction.
</td>
</tr>
</table>

To calculate the standard cell potential follow these steps:
1) Write the half-reaction and the cell potential for the reaction having the most positive (or least negative) reduction potential, E^0.
2) Write the half-reaction and the cell potential for the other reaction as an oxidation. To do this take the reaction given in the textbook table, reverse the reaction and change the E^0 sign.
3) Make sure that electrons from each half-reaction are balanced but **do not multiply the reduction potentials**.
4) Add the two half-reactions, canceling electrons, and add the two cell potentials to get the E^0_{cell}. In voltaic cells the E^0_{cell} always is a positive value indicating a spontaneous reaction.

Nernst Equation

7. What is the cell potential for the voltaic cell in Sample Exercise 5 at 325 K if the $CuSO_4$ concentration is 2.00 M and the $AgNO_3$ concentration is 3.00 M?

   The correct answer is: 0.483 V.

<table>
<tr>
<td>
E^0 is the cell potential calculated in exercise 5. R is the gas constant. n is the moles of electrons in the reaction.
</td>
<td>

$E = E^0 - \dfrac{2.303\,RT}{nF}\log Q$

$E = 0.462\ V - \dfrac{2.303\,(8.314\ J/mol\ K)(325\ K)}{(2\ mol)(9.65\times10^4\ J/V\ mol\ e^-)}\log\dfrac{2.00}{(3.00)^2}$

$E = 0.462\ V - \dfrac{6223\ V}{1.93\times10^5}\log\dfrac{2}{9}$

$E = 0.462\ V - \dfrac{6223\ V}{1.93\times10^5}(-0.653)$

$E = 0.462\ V - (-0.021\ V) = 0.483\ V$
</td>
<td>
F is faraday's constant. Q is the reaction quotient as described in module 17.
</td>
</tr>
</table>

YIELD

The cell potential determined in exercise 3 is at standard conditions (298 K, 1 M solutions, 1 atm pressure, etc.). The Nernst equation determines the cell potential under nonstandard conditions such as temperatures and solution concentrations different from standard conditions.

INSIGHT:

1) Because 1 J = V·C, $(9.65 \times 10^4 \text{ C/mol e}^-)(1 \text{ J/V·C}) = 9.65 \times 10^4 \text{ J/V mol e}^-$. Faraday's constant in a different set of units is used in this exercise.
2) This is a heterogeneous equilibrium. Solids and solutions are involved in the reaction. Equilibrium constants, including the reaction quotient Q, for heterogeneous equilibria involve species with the largest activities. Solutions have much larger activities than solids. Therefore in Q we only use solution concentrations.
3) The cell reaction is $2 \text{ Ag}^+(aq) + \text{Cu}(s) \rightarrow 2 \text{ Ag}(s) + \text{Cu}^{2+}(aq)$. Thus we get $Q = \dfrac{\left[\text{Cu}^{2+}\right]}{\left[\text{Ag}^+\right]^2} = \dfrac{2.00}{(3.00)^2} = \dfrac{2}{9}$.

Determining a Voltaic Cell Equilibrium Constant

8. What is the equilibrium constant at 298 K for the voltaic cell described in Sample Exercises 5 and 6?

The correct answer is: 4.31 x 10^{15}.

Algebra step to solve for ln K.

$$n\text{FE}^0_{cell} = RT \ln K$$

$$\frac{n\text{FE}^0_{cell}}{RT} = \ln K$$

$$\ln K = \frac{(2 \text{ mol e}^-)(9.65 \times 10^4 \text{ J/V mol e}^-)(0.462 \text{ V})}{(8.314 \text{ J/mol K})(298 \text{ K})}$$

$$\ln K = 36.0$$

$$K = e^{36.0} = 4.31 \times 10^{15}$$

All of these values were used in the previous Sample Exercises.

This cell reaction is spontaneous and very product favored. The very large K value indicates the reaction produces large product amounts.

Module 20 Predictor Questions

The following questions may help you determine the extent you need to study this module. Questions are ranked according to ability.

     Level 1 = basic proficiency
     Level 2 = mid level proficiency
     Level 3 = high proficiency

If you can correctly answer Level 3 questions you probably do not need to spend much time on this module. If you can only answer Level 1 problems, you should review this module.

Level 1   1. Calculate the mass defect, in amu, for a ^{45}Ca nucleus. The actual mass of a ^{45}Ca atom is 44.9562 amu.

Level 1   2. What is the binding energy, in J/mol, for a ^{45}Ca atom?

Level 1   3. What nuclide is the product of the alpha decay of ^{238}U?

Level 2   4. What nuclide is the product of the beta, β^-, decay of ^{137}Cs?

Level 2   5. What nuclide is the product of the positron, β^+, decay of ^{22}Na?

Level 1   6. Fill in the missing nuclide in the following nuclear reaction.

$$^{234}Th \ \rightarrow \ _{-1}^{0}\beta^- \ + \ \underline{\hspace{2cm}}$$

Level 1   7. Frequently used as a radioactive tracer in laboratory experiments, ^{32}P has a half-life of 14.28 days. If 1.50 g of ^{32}P are used in an experiment, how much is left 50.0 days later?

Level 2   8. ^{14}C, half-life of 5730 years, is frequently used to date carbon containing artifacts. If an ancient artifact has a ^{14}C content that is 20.2% that of living matter, how old is the artifact? Assume the ^{14}C decrease is entirely due to radioactive decay.

Module 20 Predictor Question Solutions

1. Calculate the mass defect, in amu, for a ^{45}Ca nucleus. The actual mass of a ^{45}Ca atom is 44.9562 amu.

   $\Delta m = [Z(1.0073) + N(1.0087) + Z(0.0005)] - actual\ mass$

   For ^{45}Ca, $Z = 20$, $N = 25$, actual mass $= 44.9562$ amu

   $\Delta m = [20(1.0073) + 25(1.0087) + 20(0.0005)] - 44.9562\ amu = 0.4173\ amu/atom$

   $$\left(\frac{0.4173\ amu}{atom}\right)\left(\frac{1.000\ g}{6.022\ x\ 10^{23}\ amu}\right)\left(\frac{60.22\ x\ 10^{23}\ atoms}{1\ mol}\right) = 0.4173\ \frac{g}{mol\ atoms}$$

2. What is the binding energy, in J/mol, for a ^{45}Ca atom?

   $\Delta m = 0.4173\ g/mol\ atoms = 4.173\ x\ 10^{-4}\ kg/mole\ atoms$

   Binding Energy $= \Delta mc^2 = (4.173\ x\ 10^{-4}\ kg/mole\ atoms)(3.00\ x\ 10^8\ m/s)^2$

   Binding Energy $= 3.7557\ x\ 10^{13}\ J/mole\ atoms$

3. What nuclide is the product of the alpha decay of ^{238}U?

   $^{238}_{92}U \rightarrow\ ^4_2He +\ ^{234}_{\underline{90}}\underline{Th}$

4. What nuclide is the product of the beta, β^-, decay of ^{137}Cs?

   $^{137}_{55}Cs \rightarrow\ ^0_{-1}\beta +\ ^{137}_{\underline{56}}\underline{Ba}$

5. What nuclide is the product of the positron, β^+, decay of ^{22}Na?

   $^{22}_{11}Na \rightarrow\ ^0_{+1}\beta +\ ^{22}_{\underline{10}}\underline{Ne}$

6. Fill in the missing nuclide in the following nuclear reaction.

   $^{234}Th \rightarrow\ ^0_{-1}\beta\ +\ \underline{\hspace{2cm}}$

   $^{234}_{90}Th \rightarrow\ ^0_{-1}\beta +\ ^{234}_{\underline{91}}\underline{Pa}$

7. Frequently used as a radioactive tracer in laboratory experiments, ^{32}P has a half-life of 14.28 days. If 1.50 g of ^{32}P are used in an experiment, how much is left 50.0 days later?

   $A = A_0 e^{-kt}$

   $$k = \frac{0.693}{t_{1/2}} = \frac{0.693}{14.28\ d} = 0.0485\ d^{-1}$$

   $A = (1.50\ g)e^{-(0.0485\ d^{-1})(50.0\ d)} = 0.133\ g$

8. ^{14}C, $t_{1/2} = 5730$ years, is frequently used to date carbon containing artifacts. If an ancient artifact has a ^{14}C content that is 20.2% that of living matter, how old is the artifact? Assume the ^{14}C decrease is entirely due to radioactive decay.

$$A = A_0 e^{-kt}$$

$$k = \frac{0.693}{t_{1/2}} = \frac{0.693}{5730 \text{ y}} = 1.21 \times 10^{-4} \text{ y}^{-1}$$

$$A = A_0 e^{-kt} \Rightarrow \ln \frac{A}{A_0} = -kt$$

$$\ln \frac{20.2}{100.} = -(1.21 \times 10^{-4} \text{ y}^{-1})t$$

$$t = 1.32 \times 10^4 \text{ y}$$

Module 20
Nuclear Chemistry

Introduction

This module discusses some basic relationships used in nuclear chemistry. The important topics described include:

1. calculating the mass defect and binding energy for a nucleus
2. predicting the products of alpha, negatron, and positron radioactive decays as well as nuclear reactions
3. problems associated with radioactive decay kinetics.

Module 20 Key Equations & Concepts

1. $\Delta m = [Z(1.0073) + N(1.0087) + Z(0.0005)]$ – **actual mass of atom**

   Nuclear mass defect is the sum of the proton, neutron, and electron masses in a nucleus minus the actual atomic mass. The mass defect describes how much nuclear mass is converted into energy to bind the nucleus.

2. **Binding Energy** $= \Delta m c^2$

   Nuclear binding energy is the energy required to bind protons and neutrons together inside the nucleus. Binding energy is the mass defect converted from mass to energy units.

3. $^A_Z X \rightarrow \, ^{A-4}_{Z-2} Y + \, ^4_2 He$

   This is the radioactive alpha decay equation. Alpha decay removes two protons and two neutrons, in the form of a ^4He nucleus, from the decaying nucleus converting element X into a new element Y.

4. $^A_Z X \rightarrow \, ^A_{Z+1} Y + \, ^0_{-1} e \quad (or \, ^0_{-1} \beta^-)$

   Radioactive beta decay, β^- or negatron decay, converts a neutron into a proton while emitting a high velocity electron, the β^- particle, from the nucleus. The decaying nucleus, X, is converted to a new nucleus, Y, having one additional proton and one less neutron than X.

5. $^A_Z X \rightarrow \, ^A_{Z-1} Y + \, ^0_{+1} e \quad (or \, ^0_{+1} \beta^+)$

   Radioactive positron decay, β^+, converts a proton into a neutron by emitting a high velocity positive electron, the β^+ particle, from the nucleus. The decaying nucleus, X is converted to a new nucleus, Y, having one less proton and one more neutron than X.

6. $^{M_1}_{Z_1} Q \rightarrow \, ^{M_2}_{Z_2} R + \, ^{M_3}_{Z_3} Y$ **where** $M_1 = M_2 + M_3$ **and** $Z_1 = Z_2 + Z_3$

   This is the nuclear reaction and radioactive decay basic relationship. Product nuclides proton numbers (Z_2 and Z_3) must sum to equal the original nuclide's proton number, Z_1. Mass numbers of the product nuclides (M_2 and M_3) must also sum to the original nuclide's mass, M_1.

7. $A = A_0 e^{-kt}$ **and** $k \, t_{1/2} = 0.693$

   Radioactive decay obeys first order kinetics as described in Module 16. These are two important radioactive decay equations describing the nuclide amount remaining after some time has passed as well as the half-life relationship.

<u>**Sample Exercises**</u>
Mass Defect
1. What is the mass defect, in amu, for a ^{55}Cr nucleus? The actual mass of a ^{55}Cr atom is 54.9408 amu.
    The correct answer is: 0.5161 amu.

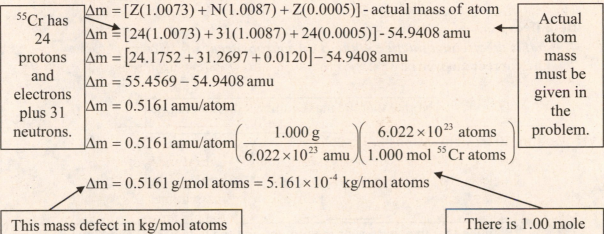

| ^{55}Cr has 24 protons and electrons plus 31 neutrons. | $\Delta m = [Z(1.0073) + N(1.0087) + Z(0.0005)]$ - actual mass of atom | Actual atom mass must be given in the problem. |

$\Delta m = [24(1.0073) + 31(1.0087) + 24(0.0005)] - 54.9408$ amu

$\Delta m = [24.1752 + 31.2697 + 0.0120] - 54.9408$ amu

$\Delta m = 55.4569 - 54.9408$ amu

$\Delta m = 0.5161$ amu/atom

$\Delta m = 0.5161 \text{ amu/atom} \left(\dfrac{1.000 \text{ g}}{6.022 \times 10^{23} \text{ amu}} \right) \left(\dfrac{6.022 \times 10^{23} \text{ atoms}}{1.000 \text{ mol } ^{55}\text{Cr atoms}} \right)$

$\Delta m = 0.5161 \text{ g/mol atoms} = 5.161 \times 10^{-4}$ kg/mol atoms

This mass defect in kg/mol atoms is used in the next exercise.

There is 1.00 mole of amu in 1.000 g.

INSIGHT: 1.0073 is the proton mass. 1.0087 amu is the neutron mass. 0.0005 amu is the electron mass.

Binding Energy
2. What is the binding energy, in J/mol, for a ^{55}Cr nucleus?
    The correct answer is: 4.64 x 10^{13} J/mol atoms.

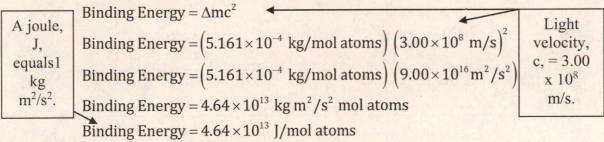

| A joule, J, equals1 kg m²/s². | Binding Energy $= \Delta mc^2$ | Light velocity, c, = 3.00 x 10^8 m/s. |

Binding Energy $= \left(5.161 \times 10^{-4} \text{ kg/mol atoms}\right) \left(3.00 \times 10^8 \text{ m/s}\right)^2$

Binding Energy $= \left(5.161 \times 10^{-4} \text{ kg/mol atoms}\right) \left(9.00 \times 10^{16} \text{ m}^2/\text{s}^2\right)$

Binding Energy $= 4.64 \times 10^{13} \text{ kg m}^2/\text{s}^2$ mol atoms

Binding Energy $= 4.64 \times 10^{13}$ J/mol atoms

Alpha Decay
3. What is the product nuclide of the alpha decay of ^{232}Th?
    The correct answer is: ^{228}Ra.

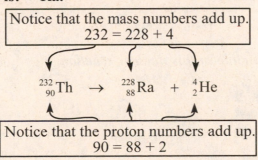

Notice that the mass numbers add up.
$232 = 228 + 4$

$^{232}_{90}\text{Th} \rightarrow {}^{228}_{88}\text{Ra} + {}^{4}_{2}\text{He}$

Notice that the proton numbers add up.
$90 = 88 + 2$

223

Beta Decay

4. What is the product nuclide of the β^-, negatron, decay of ^{14}C?
   The correct answer is: ^{14}N.

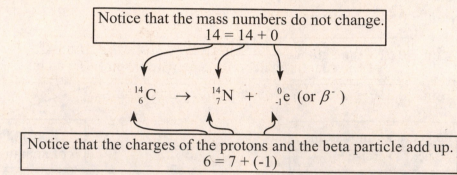

Notice that the mass numbers do not change.
$14 = 14 + 0$

$$^{14}_{6}C \rightarrow {}^{14}_{7}N + {}^{0}_{-1}e \text{ (or } \beta^- \text{)}$$

Notice that the charges of the protons and the beta particle add up.
$6 = 7 + (-1)$

5. What is the product nuclide of the β^+, positron, decay of ^{37}Ca?
   The correct answer is: ^{37}K.

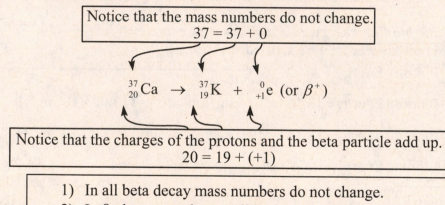

Notice that the mass numbers do not change.
$37 = 37 + 0$

$$^{37}_{20}Ca \rightarrow {}^{37}_{19}K + {}^{0}_{+1}e \text{ (or } \beta^+ \text{)}$$

Notice that the charges of the protons and the beta particle add up.
$20 = 19 + (+1)$

INSIGHT:
1) In all beta decay mass numbers do not change.
2) In β^- decay, product nuclides have one **more** proton than the decaying nuclide.
3) In β^+ decay, product nuclides have one **less** proton than the decaying nuclide.

Nuclear Reactions

6. Fill in the missing nuclide in this nuclear reaction.
$$^{53}Cr + {}^{4}He \rightarrow \underline{\hspace{1cm}} + 2n$$
The correct answer is: ^{55}Fe.

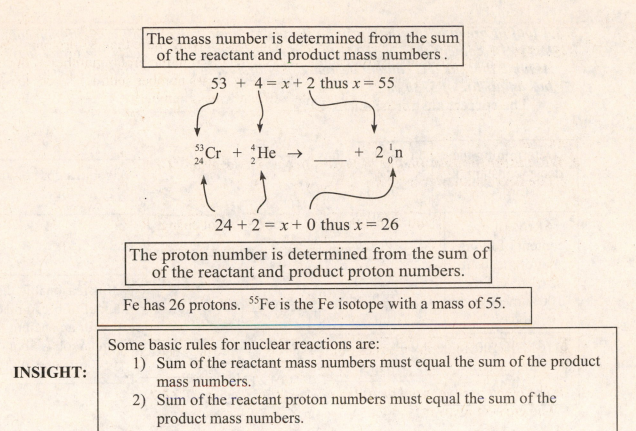

The mass number is determined from the sum of the reactant and product mass numbers.

$$53 + 4 = x + 2 \text{ thus } x = 55$$

$$^{53}_{24}\text{Cr} + {}^4_2\text{He} \rightarrow \underline{\hspace{1cm}} + 2\,{}^1_0\text{n}$$

$$24 + 2 = x + 0 \text{ thus } x = 26$$

The proton number is determined from the sum of of the reactant and product proton numbers.

Fe has 26 protons. ^{55}Fe is the Fe isotope with a mass of 55.

INSIGHT:

Some basic rules for nuclear reactions are:
1) Sum of the reactant mass numbers must equal the sum of the product mass numbers.
2) Sum of the reactant proton numbers must equal the sum of the product mass numbers.

Kinetics of Radioactive Decay

7. Tritium, ^3H, a radioactive isotope of hydrogen has a half-life of 12.26 y. If 2.0 g of ^3H were made, how much is left 18.40 y later?
   The correct answer is: 0.71 g.

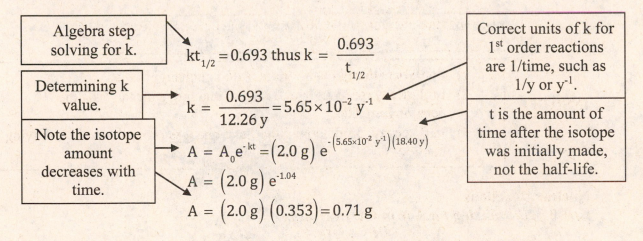

Algebra step solving for k.

$$kt_{1/2} = 0.693 \text{ thus } k = \frac{0.693}{t_{1/2}}$$

Determining k value.

$$k = \frac{0.693}{12.26\,\text{y}} = 5.65 \times 10^{-2}\,\text{y}^{-1}$$

Note the isotope amount decreases with time.

$$A = A_0 e^{-kt} = (2.0\,\text{g})\,e^{-\left(5.65 \times 10^{-2}\,\text{y}^{-1}\right)(18.40\,\text{y})}$$

$$A = (2.0\,\text{g})\,e^{-1.04}$$

$$A = (2.0\,\text{g})(0.353) = 0.71\,\text{g}$$

Correct units of k for 1st order reactions are 1/time, such as 1/y or y^{-1}.

t is the amount of time after the isotope was initially made, not the half-life.

8. *A loaf of bread left in the Egyptian temple of Mentukotep II, an ancient pharaoh, has a ^{14}C content that is 61.6% that of living matter. How old is the bread loaf? Assume that the ^{14}C content decrease is entirely due to ^{14}C radioactive decay. ^{14}C has a half-life of 5730 y.*

The correct answer is: 4.00×10^3 y.

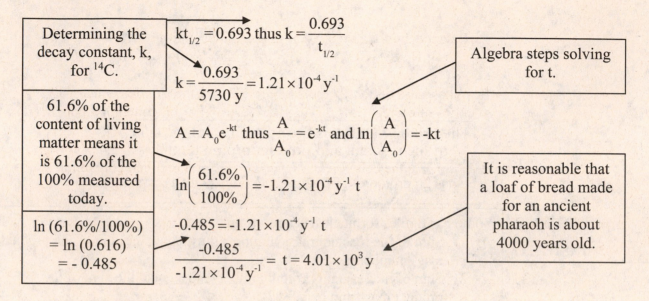

Determining the decay constant, k, for ^{14}C.

$kt_{1/2} = 0.693$ thus $k = \dfrac{0.693}{t_{1/2}}$

$k = \dfrac{0.693}{5730 \text{ y}} = 1.21 \times 10^{-4} \text{ y}^{-1}$

Algebra steps solving for t.

61.6% of the content of living matter means it is 61.6% of the 100% measured today.

$A = A_0 e^{-kt}$ thus $\dfrac{A}{A_0} = e^{-kt}$ and $\ln\left(\dfrac{A}{A_0}\right) = -kt$

$\ln\left(\dfrac{61.6\%}{100\%}\right) = -1.21 \times 10^{-4} \text{ y}^{-1} \text{ t}$

It is reasonable that a loaf of bread made for an ancient pharaoh is about 4000 years old.

$\ln(61.6\%/100\%)$
$= \ln(0.616)$
$= -0.485$

$-0.485 = -1.21 \times 10^{-4} \text{ y}^{-1} \text{ t}$

$\dfrac{-0.485}{-1.21 \times 10^{-4} \text{ y}^{-1}} = t = 4.01 \times 10^3 \text{ y}$

Practice Test Six
Modules 19-20

Level 3   1. Balance the following reaction in both acidic and basic solution.
$$Fe^{2+} + MnO_4^- \rightarrow Fe^{3+} + Mn^{2+}$$

Level 3   2. Balance the following reaction in acidic solution.
$$Br_2(l) + SO_2(g) \rightarrow Br^-(aq) + SO_4^{2-}(aq)$$

Level 1   3. What mass of copper metal is produced at the cathode when 1.30 amps of current pass through an electrolytic cell containing copper (II) sulfate for 72.0 minutes?

Level 1   4. Calculate the cell potential at 25 °C, E, for a half-cell containing Fe^{3+}/Fe^{2+} if the Fe^{2+} concentration is exactly twice that of Fe^{3+}. $E^0_{cell} = 0.771$ V

Level 1   5. Calculate the binding energy, in J/mol, for a ^{35}Cl atom. The actual mass of ^{35}Cl is 34.9689 amu.

Level 2   6. Radioactive ^{14}C, half-life of 5730 years, is frequently used to date artifacts. If an artifact has a ^{14}C content 57.4% that of living matter, how old is the artifact? Assume the ^{14}C decrease is entirely due to radioactive decay.

Level 1   7. Fill in the missing nuclide or decay particle in each nuclear reaction.

$$^{226}_{88}Ra \rightarrow \underline{\quad} + \, ^{222}_{86}Rn$$

$$\underline{\quad} \rightarrow \, ^{0}_{-1}\beta + \, ^{37}_{18}Ar$$

$$^{15}_{7}N \rightarrow \, ^{0}_{+1}\beta + \underline{\quad}$$

Math Review

Introduction

General chemistry classes require some basic mathematical skills. These include many you were taught earlier in your academic career but may have forgotten from lack of use. This section will refresh your memory on some necessary math skills for a typical general chemistry course. The important topics in this section include:

1. proper use of scientific notation
2. basic calculator skills, including entering numbers in scientific notation
3. rounding of numbers
4. use of the quadratic equation
5. the Pythagorean theorem
6. rules of logarithms

Math Review Key Equations & Concepts

1. **The quadratic equation,** $x = \dfrac{-b \pm \sqrt{b^2 - 4ac}}{2a}$

   The quadratic equation determines solutions to quadratic equations, equations of the form $ax^2 + bx + c$. Quadratic equations frequently occur in equilibrium problems.

2. **Pythagorean theorem,** $a^2 + b^2 = c^2$

   Pythagorean theorem determines the length of one side of a right triangle given lengths of the other two sides. It is used to determine the edge length or the diagonal length of a cubic unit cell in the section when calculating atomic or ionic radii.

3. $x = a^y$ **then** $y = \log_a x$

   $\log(x \cdot y) = \log x + \log y$

   $\log\left(\dfrac{x}{y}\right) = \log x - \log y$

   $\log(x^n) = n \log x$

   The first equation defines logarithms. Other equations are basic algebra rules involving logarithms. These rules apply to logarithms of any base, including base e or natural logarithms, ln. Logarithms are frequently used in kinetics and thermodynamic expressions.

Scientific and Engineering Notation

In the physical and biological sciences it is frequently necessary to write extremely large or small numbers. It is not unusual for these numbers to have 20 or more digits beyond the decimal point. For the sake of simplicity and to save space when writing, a compact or shorthand method of writing these numbers is employed. There are two possible but equivalent methods called either scientific or engineering notation. In both methods the insignificant digits serving as placeholders between the decimal place and the significant figures are expressed as powers of ten. Significant digits are then multiplied by the appropriate powers of ten to give a number that is both mathematically correct and indicative of the correct number of significant figures for the problem. To be strictly correct, the significant figures should be between 1.000 and 9.999; however, this particular rule is frequently ignored. In fact, it must be ignored when adding numbers in scientific notation having different powers of ten.

The major difference between scientific and engineering notation is how the powers of ten are written. Scientific notation uses the symbolism "x 10^y" whereas engineering notation uses the symbolism "Ey" or "ey". Engineering notation frequently is used in calculators and computers.

INSIGHT: *Positive powers of ten* indicate the decimal place has been *moved to the left that number of spaces*.
Negative powers of ten indicate the decimal place has been *moved to the right that number of spaces*.

A few examples of both scientific and engineering notation are given in this table.

Number	Scientific Notation	Engineering Notation
10,000	1 x 10^4	1E4
100	1 x 10^2	1E2
1	1 x 10^0	1E0
0.01	1 x 10^{-2}	1E-2
0.000001	1 x 10^{-6}	1E-6
23,560	2.356 x 10^4	2.356E4
0.0000965	9.65 x 10^{-5}	9.65E-5

For your chemistry success you have to understand using both methods of expressing very large or small numbers. Familiarize yourself with both methods.

Basic Calculator Skills
General Chemistry courses require calculations that are frequently performed on calculators. You do not need to purchase an expensive calculator for your course. Instead, you need a calculator that has some basic function keys. Common important functions to look for on a scientific calculator are: log and ln, antilogs or 10^x and e^x, ability to enter numbers in scientific or engineering notation, x^2, $1/x$, $\sqrt{}$ or multiple roots, like a cube or higher root.

More important than having an expensive calculator is knowing how to use your calculator. It is strongly recommended that you study the manual that comes with your calculator to learn some basic skills of entering numbers and understanding answers that your calculator provides. For a typical general chemistry course there are three important calculator skills in which you must be proficient.

1) <u>Entering Numbers in Scientific Notation</u>

Get your calculator and enter into it the number 2.54×10^5. The correct sequence of strokes is: press 2, press the decimal button, press 5, press 4, and *then press either EE, EX, EXP or the appropriate exponential button on your calculator*. **Do not press x 10 before you press the exponential button!** That's a very common mistake that will cause your answer to be 10 times too large.

After you have entered 2.54×10^5 into your calculator, press the Enter or = button and look at the number display. *If it displays 2.54E6 or 2.54 x 10^6, you have mistakenly entered the number*. Correct your number entering procedure early in the course before it becomes a bad habit!

2) <u>Taking Roots of Numbers and Entering Powers</u>

Frequently we must take a square or cube root of a number to answer a problem. Most calculators have a square root button, $\sqrt{}$. To take a square root, simply enter the number into your calculator and press the $\sqrt{}$ button to get your answer. For example, take the square root of 72 (the answer is 8.49).

Some calculators have a $\sqrt[3]{}$ button as well. If your calculator does not have a $\sqrt[3]{}$ button, then you can use the y^x button to achieve the same result. To take a cube root, enter 1/3 or 0.333 as the power and the calculator will take a cube root for you. For example, enter $27^{0.333}$ into your calculator (the correct answer is 3.00). If you need a fourth root, enter ¼ (0.25) as the power. Higher roots are determined similarly.

3) <u>Taking base 10 logs and natural logs, ln</u>

Many thermodynamic, equilibrium, and kinetic functions require the use of logarithms. All scientific calculators have log and ln buttons. To use them simply enter your number and press the button. For example, the log 1000 = 3.00, and the ln of 1000 = 6.91.

INSIGHT: A common mistake is calculating ln when log is needed and vice versa. Be careful which logarithm you calculate for the problem.

Rounding Numbers

When determining the correct number of significant figures for a problem it is frequently necessary to round off an answer. Basically, if the number immediately after the last significant figure is a 4 or lower, round down. If it is a 6 or higher, round up. The confusion arrives when the determining number is a 5. If the following number is a 5 followed by a number greater than zero, round the number up. If the number after the 5 is a zero, then the textbook used in your course will have a rule based upon whether the following number is odd or even. You should use that rule to be consistent with your instructor. The following examples illustrate these ideas. In each case the final answer contains three significant figures.

Initial Number	Rounded Number
3.67492	3.67
3.67623	3.68
3.67510	3.68
3.67502	Use your textbook rule.

Solving Quadratic Equations

Equilibrium problems frequently require solutions to quadratic equations having the form $ax^2 + bx + c$. Both solutions are determined using this formula.

$$x = \frac{-b \pm \sqrt{b^2 - 4ac}}{2a}$$

For example, if the quadratic equation is $3x^2 + 12x - 6$, then a = 3, b = 12, and c = -6. The two solutions are found as follows.

$$x = \frac{-b \pm \sqrt{b^2 - 4ac}}{2a}$$

$$x = \frac{-12 \pm \sqrt{12^2 - 4(3)(-6)}}{2(3)}$$

$$x = \frac{-12 \pm \sqrt{144 + 72}}{2(3)}$$

$$x = \frac{-12 \pm \sqrt{216}}{6}$$

$$x = \frac{-12 \pm 14.7}{6} = \frac{2.7}{6} \text{ and } \frac{-26.7}{6}$$

$$x = 0.45 \text{ and } -4.45$$

INSIGHT: Quadratic equations always have two solutions. In equilibrium problems, one of the solutions will not make physical sense. For example, one solution gives a negative solution concentration or a concentration outside the possible solution concentration ranges. Based on your problem knowledge you must choose the correct solution.

The Pythagorean Theorem

In Module 13 we determined the radius of an atom in a cubic unit cell. Because the cell is cubic, a right triangle can always be formed using two sides and the face diagonal. The face diagonal length can be determined using the Pythagorean theorem. A unit cell geometry example and determination of face diagonal length is given below.

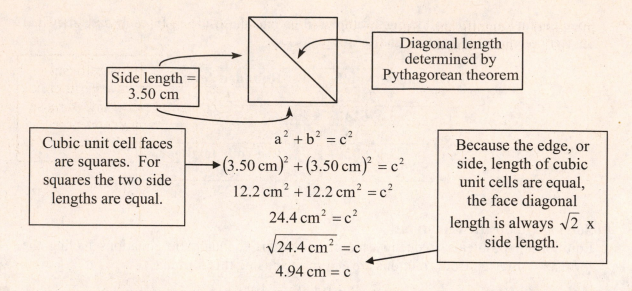

$$a^2 + b^2 = c^2$$
$$(3.50 \text{ cm})^2 + (3.50 \text{ cm})^2 = c^2$$
$$12.2 \text{ cm}^2 + 12.2 \text{ cm}^2 = c^2$$
$$24.4 \text{ cm}^2 = c^2$$
$$\sqrt{24.4 \text{ cm}^2} = c$$
$$4.94 \text{ cm} = c$$

Side length = 3.50 cm

Diagonal length determined by Pythagorean theorem

Cubic unit cell faces are squares. For squares the two side lengths are equal.

Because the edge, or side, length of cubic unit cells are equal, the face diagonal length is always $\sqrt{2}$ x side length.

Logarithm Rules

Logarithms are convenient methods to write exceptionally large or small numbers and express exponential functions. Logarithms also have the convenience factor of making multiplication and division of numbers written in scientific notation especially easy. In logarithmic form addition and subtraction of numbers is all that is required. By definition, a logarithm is the number that the base must be exponentially raised to in order to produce the original number. For example, if the number we are working with is 1000 then 10, the base, must be cubed, raised to the 3rd power, to reproduce it. Mathematically, we are stating that $1000 = 10^3$, so the log (1000) = 3. There are four commonly used logarithm rules you must know. They are given below.

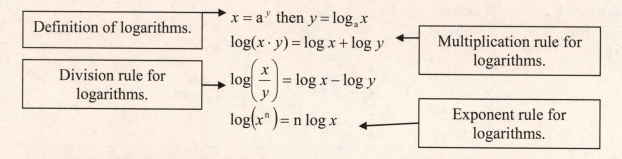

Definition of logarithms.

$$x = a^y \text{ then } y = \log_a x$$
$$\log(x \cdot y) = \log x + \log y$$
$$\log\left(\frac{x}{y}\right) = \log x - \log y$$
$$\log(x^n) = n \log x$$

Multiplication rule for logarithms.

Division rule for logarithms.

Exponent rule for logarithms.

INSIGHT: These rules are correct for base 10, natural, or any other base logarithms.

Significant Figures for Logarithms

There are 4 significant figures in the number 2.345×10^{12} (the 2, 3, 4, and 5). The power of 10 (the number 12) is not significant. If we take the log of 2.345×10^{12} the number of significant figures must remain the same. The log of $2.345 \times 10^{12} = 12.3701$.

In logarithms what numbers express the exponents present in scientific notation? They are the numbers to the left of the decimal place. The numbers to the left of the decimal place in logarithms (the characteristic) are insignificant and the ones to the right (the mantissa) are significant. Knowing this we can write correctly the log (2.345×10^{12}) = 12.3701 giving both numbers 4 significant figures.

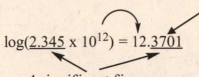

$$\log(\underline{2.345} \times 10^{12}) = 12.\underline{3701}$$

4 significant figures

12 is insignificant because in logarithms it serves the same purpose as an exponent in scientific notation.

Practice Test One Solutions

1. 5.31×10^6 dm^3

2. 1.468×10^5
   27.340 (5 sig fig)
   6.00 (3 sig fig)
   $27.340 - 6.00 = 21.34$ (4 sig fig)
   6.8371×10^3 (5 sig fig)
   $21.34 \times 6.8371 \times 10^3 = 1.459 \times 10^5$ (4 sig fig)
   871.4 (4 sig fig)
   $1.459 \times 10^5 + 871.4 = 1.468 \times 10^5$ (4 sig fig)

3. 5.12 g

4. 4.64×10^{24} atoms of N

5. 1.817 mol $Ca_3(AsO_4)_2$

6. a) 2 iron (III) ions
   b) 3 sulfate ions
   c) 0 sulfide ions
   d) 0 oxide ions

7. a) phosphorus pentachloride
   b) ammonium sulfate
   c) lithium nitrate
   d) potassium dihydrogen borite
   e) xenon tetrafluoride

8. a) SF_6
   b) $HCN(aq)$
   c) $Cu(OH)Cl$
   d) $MgBr_2$
   e) $HClO$

Practice Test Two Solutions

1. 22.8 g oxygen

2. $\underline{1}\ P_4O_{10} + \underline{6}\ H_2O \rightarrow \underline{4}\ H_3PO_4$

3. $AgNO_3$ is the limiting reagent. The theoretical yield of $Ca(NO_3)_2$ is 13.72 g. The percent yield is 78.11%.

4. 13.6 mol HCl

5. The required volume is 441 mL.

6. The final volume is 130. mL.

7. $2H_2O \rightarrow 2H_2 + O_2$ decomposition and oxidation/reduction
   $H_2 + Cl_2 \rightarrow 2HCl$ combination and oxidation/reduction
   $AlCl_3 + 3AgNO_3 \rightarrow 3AgCl + Al(NO_3)_3$ metathesis and precipitation

8. Total ionic equation: $2H^+ + 2\ ClO_3^- + Sr^{2+} + 2\ OH^- \rightarrow Sr^{2+} + 2\ ClO_3^- + 2H_2O$
   Net ionic equation: $H^+ + OH^- \rightarrow H_2O$ or $2\ H^+ + 2\ OH^- \rightarrow 2\ H_2O$

Practice Test Three Solutions

1. $1s^2 2s^2 2p^6 3s^2 3p^6 4s^2 3d^8$ or $[Ar]\ 4s^2 3d^8$
   a) n = 3
   b) 6 paired electrons
   c) 2 unpaired electrons
   d) $\ell = 2$

2. a) 10
   b) 2
   c) 1
   d) 6
   e) 2

3. Oxygen is the most electronegative element. Nitrogen has the highest first ionization energy.

4. Cl releases the most energy upon accepting an electron. Its electron affinity is very negative.

5. Atomic radii increase down a group because of the increase in principle quantum number. Electrons are farther away from the nucleus. From left to right across a period, the principle quantum number is the same; however, effective nuclear charge increases due to decreased shielding. The increase in effective nuclear charge pulls electrons closer to the nucleus, resulting in a smaller radius.

6. MgO is an ionic compound (metal bonded to a nonmetal). There are two ions, Mg^{2+} and O^{2-}.

$$[Mg]^{2+}\ [:\overset{..}{O}:]^{2-}$$

7. SF₄ is a covalent compound (two nonmetals bonded together). There are no ions present. In the Lewis structure, S is the central atom with four single bonds to the four F atoms. There is also a lone pair on the central S atom.

8. XeF₄   octahedral electronic geometry; square planar molecular shape
   I₃⁻   trigonal bipyramidal electronic geometry; linear molecular shape
   CO₂   linear electronic geometry; linear molecular shape
   C₂H₄   trigonal planar electronic geometry; trigonal planar molecular shape about both C atoms

9. All of the molecules and ions in question 8 are non polar.

10. XeF₄   sp³d²
    I₃⁻   sp³d
    CO₂   sp
    C₂H₄   sp² for both C atoms

Practice Test Four Solutions

1. Strong acids: HNO₃, HCl, HI, HBr, H₂SO₄, HClO₃ (HClO₄ is also correct.)
   Strong bases: LiOH, NaOH, KOH, RbOH, CsOH, Sr(OH)₂, Ba(OH)₂

2. The true statements are a), c), and e).

3. a) CH₄   London dispersion
   b) CH₂Cl₂   diple-dipole
   c) CH₃COOH   hydrogen bonding
   d) HF   hydrogen bonding
   e) PCl₃   dipole-dipole

4. CaO (3850 °C) > CH₃COOH (118 °C) > CH₂Br₂ (96.95 °C) > CCl₄ (76.72 °C)

5. V = 49.6 L

6. There are two atoms per unit cell, so it is a body-centered cubic unit cell.

7. % w/w = 21.2%
   $X_{H_3PO_4} = 0.0472$

8. 34.5 g/mol

9. 7.03×10^4 J

10. ΔH^0_{rxn} = -1516.8 J The negative sign indicates the reaction is exothermic.

Practice Test Five Solutions

1. rate = $k[A]^2[C]$

2. The reaction is first order (from the units of the rate constant); $[N_2O_5]$ = 0.160 M

3. t = 7.92×10^3 s or 132 min

4. E_a = 3.0×10^4 J/mol

5. K_c = 8.0×10^{-12}

6. $[PCl_3]$ = 0.071 M

7. Since the reaction is endothermic, heat is a reactant. Increasing the temperature is the equivalent of adding reactant. Addition of a reactant shifts equilibrium toward the products.

8. pH = 7.26

9. pOH = 6.13; pH = 7.87

10. pH = 3.90

Practice Test Six Solutions

1. Acidic solution: $5\ Fe^{2+} + MnO_4^- + 8\ H^+ \rightarrow 5\ Fe^{3+} + Mn^{2+} + 4\ H_2O$
   Basic solution: $5\ Fe^{2+} + MnO_4^- + 4\ H_2O \rightarrow 5\ Fe^{3+} + Mn^{2+} + 8\ OH^-$

2. $Br_2 + 2\ H_2O + SO_2 \rightarrow 2\ Br^- + SO_4^{2-} + 4\ H^+$

3. 1.85 g $Cu_{(s)}$

4. E = 0.753 V

5. BE = 2.8827×10^{13} J/mol atoms

6. 4.59×10^3 y

$^{4}_{2}\text{He}$

7. $^{37}_{17}\text{Cl}$

$^{15}_{6}\text{C}$